CEREBRUM 2013

Cerebrum 2013
Emerging Ideas in Brain Science

Bill Glovin, Editor

D
DANA
PRESS

New York

Published by Dana Press, a Division of the Dana Foundation, Incorporated

Address correspondence to:
Dana Press
505 Fifth Avenue, Sixth Floor
New York, NY 10017

THE
DANA
FOUNDATION
New York, NY 10017

DANA is a federally registered trademark.

Printed in the United States of America

ISBN-13: 978-1-932594-61-4
ISSN: 1524-6205

Book design by Bruce Hanson at EGADS (egadsontheweb.com)
Cover illustration by Danny Schwartz

CONTENTS

CEREBRUM 2013

Foreword

By Bruce S. McEwen, Ph.D.

Bruce S. McEwen, Ph.D., is the Alfred E. Mirsky Professor and head of the Harold and Margaret Milliken Hatch Laboratory of Neuroendocrinology at the Rockefeller University. He is a member of the National Academy of Sciences, the Institute of Medicine, and the American Academy of Arts and Sciences, and a fellow of the New York Academy of Sciences. He is a past president of the Society for Neuroscience. McEwen's laboratory discovered adrenal steroid receptors in the hippocampus in 1968 and, later, sex hormones in this brain region. His current research focuses on the effects of stress on the amygdala, prefrontal cortex, and hippocampus, along with sex-hormone effects and sex differences in these brain regions. He is also involved in social neuroscience because of the brain's central role in human adaptation to our changing and often stressful world. McEwen is co-author of *The End of Stress as We Know It* (Dana Press/Joseph Henry Press, 2002) and *The Hostage Brain* (The Rockefeller University Press, 1994).

NEUROSCIENCE IS AN AMAZINGLY broad and diverse field, ranging from fundamental studies of basic brain processes such as language, learning, and memory, to brain-body interactions and the neurochemical and hormonal basis of behaviors, to pathophysiological changes in neurodegenerative disease and mental-health disorders. Neuroscience also includes such topics as bioethics and the emerging field of social neuroscience.

My own evolution as a neuroscientist and a social neuroscientist (sometimes I call myself a "molecular sociologist") began with finding receptors for adrenal steroids and, later, for sex hormones in the hippocampus, a brain region involved in memory and mood regulation. These hormones and their receptors are very important in connecting us to the social and physical environment and the experiences of everyday life, in which the brain is the central organ that keeps us in touch with both the outside world and our internal milieu. My colleagues and I used these hormones and their receptors to demonstrate structural plasticity in terms of synapse turnover, dendritic remodeling, and neurogenesis in brain regions involved in higher cognitive functions and mood regulation in both adult and developing brains, accompanied by relevant behavioral changes (e.g., alterations in memory, mood, decision making, and cognitive flexibility). Now we investigate how hormones and experiences alter gene expression via epigenetic mechanisms to produce these structural and behavioral effects in those same brain regions.

In parallel with my work as head of a research laboratory, I participate in two interdisciplinary networks—the MacArthur Research Network on Socioeconomic Status (SES) and Health, and the National Scientific Council on the Developing Child—in which I help apply neuroscience and stress biology to understand human health issues such as gradients of common disorders like obesity, diabetes, cardiovascular disease, depression, and certain cancers across income and education, as well as how early-life adversity "gets under the skin" and causes lifelong increases in susceptibility to these disorders. Although the causes of these SES gradients of health are very complex, they are likely to reflect, with increasing frequency at the lower end of the scale, the cumulative burden of coping with limited resources, poor education, and negative life events, as well as differences in lifestyle and resulting chronic activation of physiological systems involved in adaptation.

Realizing the ambiguity and inadequacy of the term "stress" and recognizing that a person's daily experiences as a social creature and his or her individual health behaviors (lifestyle) operate through the brain to regulate neuroendocrine, autonomic, metabolic, and immune body systems, I helped develop a concept called allostatic load. Allostatic load, or its more extreme form, allostatic overload, refers to the fact that activation of these body systems promotes adaptation to daily challenges (via the process of allostasis, meaning "maintaining homeostasis through active change"), but that too much stress and dysregulation of the individual adaptive body systems with respect to one another can lead to physical and mental wear and tear that can promote common disorders. An example of allostatic load is an increase in body fat. When this happens in a bear preparing for winter, the fat provides an energy source during hibernation. However, for a human being whose lifestyle involves long work hours every day, poor sleep, lack of exercise, and a diet of comfort foods—or for a bear that lives a boring, sedentary life in a zoo—an increase in abdominal fat causes an elevated systemic inflammatory burden that contributes to cardiovascular disease and diabetes. The concepts of allostatic load and overload have become useful in creating a bridge between brain and body and among the fields of neuroscience, medicine, psychiatry, epidemiology and public health, psychology, and sociology.

Combined with the concepts of allostatic load and overload, two realizations—that the entire brain shows adaptive plasticity and is a malleable target of steroid and metabolic hormones and nutrients, and that the brain is the central organ of stress and adaptation to a changing social and physical environment—have created a new way of understanding what determines human health and longevity. Moreover, this new view opens possibilities for improving health at different levels. These possibilities include informing government and private-sector policies on education, work commuting, recreation, and support for families, as well as helping individuals handle the challenges of modern life. Behavioral interventions will play a large role; pharmacological agents will be adjuncts rather than final solutions. Indeed, social neuroscience has the potential to impact the lives of millions of people.

With this in mind, let us consider other topics covered in this anthology of 2013 *Cerebrum* stories: up-to-date overviews of brain disorders, behavioral disorders, and the processes of learning and exploring the world; new insights into key brain-body interactions; the unfortunate stalemate in developing new psychiatric drugs; and ongoing concerns about scientific misconduct that reflect the current social environment of science.

The Hostage Brain

The brain is hostage to disorders that arise internally but are often precipitated by or exacerbated by injury or stressful events. Rather than being held hostage to a lack of knowledge, researchers are progressing in a number of interesting new ways, illuminating targets for both treatment prevention, and revealing the impact of the social environment on various disorders. Migraine is a serious problem that affects not only the brain and behavior, but also systemic physiology via inflammation and insulin resistance. In this volume, Andrew H. Ahn and Peter J. Goadsby describe new research uncovering an important link between migraine and sleep patterns. Their work holds enormous promise of improved care for millions of people who experience migraine and suffer from familial advanced sleep phase syndrome (FASP).

Alzheimer's disease may be the most familiar type of dementia, but Lewy body dementia (LBD) is actually the most prevalent progressive dementia. LBD is characterized by the presence of Lewy bodies, which are abnormal aggregates of a protein called alpha-synuclein. They are found in brain regions that regulate behavior, memory, movement, and personality. In their article, James E. Galvin and Meera Balasubramaniam point out that many of the symptoms of Alzheimer's disease, Parkinson's disease, and LBD overlap, but LBD is more difficult to diagnose. Underdiagnosis is just part of the reason why LBD is unknown to both the public and many health-care providers. It also explains why funding for LBD research lags far behind that for almost every other cognitive disorder.

External causes of brain damage include accidents, military injuries, and concussions related to culturally important contact sports such as football and hockey. Chronic traumatic encephalopathy (CTE) is a long-term, degenerative, incurable brain disease caused by repeated hits to the head. While scientists have linked concussions to brain and central nervous system issues for a long time, a new study suggests that repeated hits to the head—mild or otherwise—can lead to memory loss, depression, and dementia. This postmortem brain study, conducted at the Boston University Center for the Study of Traumatic Encephalopathy, provides new and troubling evidence about CTE. Although military personnel and others are vulnerable to the disease, the highest risk is among athletes involved in contact sports in which hits to the head are considered part of the game. Chris Nowinski, a former college football player and professional wrestler, writes about how a concussion put him on the path of dedicating his life to making others aware of CTE's dangers and helping develop a treatment.

While the brain is normally protected from autoimmune disorders, there are exceptions, as in multiple sclerosis and the paraneoplastic diseases. In *Cerebrum 2013*, David Lynch reviews the book *Brain on Fire: My Month of Madness* by Susannah Cahalan, who, in a medical detective story, describes her experience with the rare disorder anti-NMDA-receptor encephalitis, which results from the production of an autoimmune response against the N-methyl-D-aspartate (NMDA) receptor. The NMDA receptor regulates synaptic function and plasticity in the brain and is critical for learning and memory. For Cahalan, the syndrome evolved over several weeks and presented as a psychosis similar to that seen in schizophrenia. At the book's beginning, she is a young reporter struggling to find story ideas that will resonate with her editors at the *New York Post*. Over several weeks, she gradually finds her personality changing as she begins suffering from paranoia, hallucinations, and seizures and withdraws from her friends and colleagues.

Epilepsy affects nearly 3 million Americans of all ages. The incidence of epilepsy is greater in African-Americans and in other socially disadvantaged populations, and about 200,000 new cases are diagnosed each year. While drugs work for some, others find them ineffective. What seems to work just as well, if not better, especially in children, is the ketogenic diet, a relatively unknown high-fat diet. In "Epilepsy's Big Fat Answer," John M. Freeman, one of the nation's leading advocates for the diet's use, writes about the evolution of the ketogenic diet and people's resistance to accepting it. One of the first and most successful therapies for epilepsy, the ketogenic diet fell into obscurity with the introduction of anticonvulsant medications in the 1930s. Today, however, the diet has reemerged as a therapeutic option, and scientists are exploring its effectiveness for other neurological disorders, including brain tumors, autism, and even Alzheimer's disease. The diet produces ketones, the residues left when fats are burned in the absence of sufficient glucose. Ketones are, in fact, an efficient energy source for the brain, and for unknown reasons, they make the ketogenic diet—high in fat, low in carbohydrates—more effective than current anticonvulsant medications in curbing difficult-to-control seizures.

Reciprocal Communication Between Brain and Body

Just as diet has important effects upon the nervous system, the brain and body communicate reciprocally, and chemicals produced by bacteria as well as the immune system influence brain function and mood. One of the brain and body chemicals that mediate these effects are the endogenous cannabinoids (endocannabinoids). Bradley E. Alger reviews the endocannabinoid system—named after the plant that led to its discovery—which provides mechanisms that connect brain activity and body function in health and disease. Endocannabinoids and their receptors, which are found throughout the body and brain, interact with our organs, connective tissues, glands, and immune cells to produce obesity in the body and to regulate states of anxiety in the brain.

There is growing recognition of the influence of the microbiome in our gut as a factor in brain-body communication, now referred to as the gut-brain axis. Microbiota in our gut, sometimes referred to as the "second genome" or the "second brain", may influence our mood, as well as our systemic physiology, in ways that scientists are just now beginning to understand, as addressed by Jane A. Foster. As research evolves from mice to people, a further understanding of microbiota's relationship to the human brain could have significant mental-health implications. It could also improve treatment of GI disorders such as *C. difficile* infection through fecal transplants, which, in turn, may affect brain function and mood, among other results.

Inflammation is part of the innate immune response. It is also a widespread process throughout the body that is associated with most diseases of modern life—including diabetes, cardiovascular disease, arthritis, cancer, and dementia—and, now, with linkages to depression and early-life adversity. Many tissues, including the brain, generate pro- and anti-inflammatory cytokines. Charles L. Raison and Andrew H. Miller examine what the latest research reveals about the link between inflammation in the brain and depression, and explain how a better understanding of that link can be a critical first step in the personalization of care and treatment of mood disorders.

The Absence of New Psychiatric Drugs

In spite of the increasing prevalence of mental-health disorders worldwide, the pharmaceutical industry has reduced funding for disorders such as depression, anxiety disorders, and schizophrenia. Instead, it relies on existing drugs, with their limitations, until neuroscientists make new advances in this area. Steven E. Hyman discusses this paradox in terms of the need for researchers to provide better insights into the genetics and epigenetics of brain function and malfunction, which, he says, will be the role of academia, rather than the pharmaceutical industry, in the near future. He notes the recognition by ongoing researchers of defects in neuronal architecture as a basis for mental-health disorders and the need to better understand and manipulate the brain's potential for plasticity, including possible uses of engineered stem cells.

Scientific Misconduct

The current atmosphere of science worldwide is filled not only with exciting possibilities for discovery, but also with increased competition over grant money, placement of articles in prestigious journals, and academic appointments and advancement. Judging by a tenfold increase in retractions of articles over the past decade, fraud, or scientific misconduct, is on the rise. Scientific misconduct includes plagiarism, faked data, and altered images in publications. Stephen G. Lisberger, an experienced editor and distinguished neuroscientist, provides a thoughtful discussion of this problem and its potential solutions.

New Insights into Behavior and Brain Development

Modern neuroscientists are providing insights into brain mechanisms for risk-taking, language and literacy, children's ability to succeed, and brain function altered by attention-deficit/hyperactivity disorder (ADHD) and autism. Lee Alan Dugatkin tackles the question of understanding individual differences in risk-taking, or what he calls boldness, in animals. He suggests that knowledge about animal boldness will not only help humans understand and live more easily with creatures of other species, but also inform our own risk-taking tendencies.

ADHD is a troubling aspect of human brain development and resulting behavior. According to Philip Shaw, modern brain-imaging methods are providing new insights into the basis of ADHD. Given these insights, along with the growing realization that the brain's architecture continues to be plastic throughout the life course, scientists hope to develop better methods of controlling and eventually treating ADHD in both children and adults.

Autism is another developmentally related condition of the human brain that impairs social interactions. Robert L. Findling reviews *The Autistic Brain: Thinking Across the Spectrum*, a new book by Temple Grandin and Richard Panek. The book is an up-to-date overview of advances in the study of the autistic brain, as well as an insightful look at Grandin's own remarkable progression to become a professor of animal science at Colorado State University.

The ability to read, so vital to modern human life, involves a very specific area of the brain that Stanislas Dehaene has referred to as the letter box: a visual cortical area next to the area for spoken language. This area, Dehaene says, has undergone so-called neuronal recycling to acquire new functions and connections among cortical areas in the 4,000 years of written language. His research on illiterate versus literate people shows the degree of remodeling of these cortical areas that occurs with the acquisition of literacy.

Going beyond reading to understand how children succeed in modern society, Silvia A. Bunge reviews Paul Tough's book, *How Children Succeed: Grit, Curiosity, and the Hidden Power of Character.* (This title follows Tough's first book, *Whatever It Takes: Geoffrey Canada's Quest to Change Harlem and America*, which discusses the creation of the Harlem Children's Zone, Inc.) The reference to character in the book's title alludes to good self-control, which, Bunge notes, ". . . encompasses layman's terms like 'willpower,' 'character,' and 'grit,' as well as technical terms like 'executive functions,' 'cognitive control processes,' or 'emotion regulatory strategies.' . . . Self-control makes it possible to sit still, listen quietly, keep relevant rules and goals in mind, work through feelings rather than erupting in anger, and consider long-term consequences before acting. As such, it makes good sense that research points to self-control as essential for scholastic achievement and good life outcomes." Self-control involves healthy development of brain architecture involving, in particular, the prefrontal cortex, which matures during childhood and into young adulthood and is thus sensitive to and malleable by the quality of the social environment. Overall, Bunge says, Tough's book "makes a powerful argument for giving children opportunities to rise out of poverty." Here, improving the social environment of children and their families, minimizing adverse childhood experiences, and stimulating language development will be very important.

ARTICLES

1

The Evolution
of Risk-Taking

By Lee Alan Dugatkin, Ph.D.

Lee Alan Dugatkin, Ph.D., is a professor and Distinguished University Scholar in the Department of Biology at the University of Louisville. His main area of research interest is the evolution of social behavior. He is currently studying the evolution of cooperation, the evolution of aggression, the interaction between genetic and cultural evolution, the evolution of antibiotic resistance, the evolution of senescence, and the evolution of risk-taking behavior. Dugatkin has authored *Cooperation among Animals: An Evolutionary Perspective, Cheating Monkeys and Citizen Bees*, and *The Altruism Equation: Seven Scientists Search for the Origins of Goodness*. Dugatkin is also the author of two textbooks: *Principles of Animal Behavior* and, along with coauthor Carl Bergstrom, *Evolution*.

Many animal species besides humans show evidence of individuality. Knowing how a risk-taker differs from its stay-at-home counterpart could not only help humans live more easily with our fellow creatures, says Lee Dugatkin of the University of Louisville, but also tell us a few things about ourselves and how we got this way.

I'M GOING TO MAKE THE CASE that we can study the evolution of risk-taking by closely examining work on this subject in nonhumans. Researchers have found several ways to study behavior and emotion in animals. Sound fishy? Here's a beautiful example of how it is done.

Psychologists use a suite of measures to assess what they call "subjective well-being": A composite well-being score is assigned to you after others rate you on how happy you appear to be. Ethologists, scientists who study animal behavior, recently modified the scale to apply it to orangutans. Alexander Weiss and his team found two general types of orangutans residing in zoo populations. One type, which seemed happy and outgoing, interacted in positive ways with other apes and with their zookeepers and rarely showed abnormal or neurotic behavior. The second type showed the opposite set of traits (Weiss et al. 2006).

So far, so good. But does what Weiss's team found translate in any real way to what we mean by the common usage of "happy and well-adjusted"? It does in orangutans: The apes that the Weiss group scored high on the well-being scale lived far longer. Orangutans that were one standard deviation above the group norm for subjective well-being lived about 11 years longer than did their subjectively less happy group mates, which were one standard deviation below the group mean.

If we can study subjective well-being in nonhumans then, in principle, we can study risk-taking as well. And there is every reason we'd want to do that. In humans, risk-taking, often referred to as boldness, is one of the most consistent personality traits displayed over the course of an individual's lifetime. The same is true for the other end of the continuum, risk aversion (which overlaps, but is not synonymous, with shyness). And psychologists, perhaps most notably Jerome Kagan, have done wonderful work on risk-taking and shyness in humans. But there are limitations (Kagan 1994). Ethologists point out that, from an evolutionary perspective, research on humans is lacking for at least two reasons. First, at the species level, humans are a sample size of one. Second, we don't know much about the costs and benefits of risk-taking in natural settings, but it is knowledge of the costs and benefits of a behavior that allows us to use natural-selection thinking to understand evolutionary history and to make predictions about the future. We could remedy both problems by looking at the costs and benefits of risk-taking across many species. And that is what we have started to do.

Curious Fish

One of the earliest controlled ethological studies of risk-taking began in the experimental ponds at Cornell University, where David Sloan Wilson and his colleagues began searching for different personality types in the pumpkinseed sunfish (*Lepomis gibbosus*). Based on work in humans, Wilson and his team hypothesized that risk-taking sunfish would be especially likely to explore novel objects and novel environments (Wilson et al. 1993; Wilson et al. 1994). They collected fish from a pond with a series of funnel-shaped traps designed so that if a fish swam into one, it would be very hard to swim back out. Presumably, risk-taking fish would be more likely to explore such a novel trap, and the researchers would catch more risk-taking fish than risk-averse fish. The researchers also dragged a large net through the pond and scooped up all the fish they could, both bold and shy. If Wilson and his team were correct, the fish from the funnel traps would be bolder on average than those from the netted samples.

Wilson's team found that pumpkinseed populations are mixtures of bold and shy fish. For example, stomach-content analysis indicated that fish caught in the funnel traps had been eating more often and in areas of the pond that were the least safe—the sorts of behaviors that we might expect from risk-taking individuals. Funnel-trapped fish also had many more parasites than did netted fish, suggesting that they explored many areas of their pond and thus exposed themselves to myriad parasites.

When funnel-trapped and netted fish were marked individually with colored beads and released back into their pond, funnel-trapped fish were much more likely to be found leaving their group and foraging on their own than were net-trapped fish. And, finally, when pumpkinseed fish were brought into the laboratory and exposed to a novel object there, funnel-trapped individuals were much more likely to investigate the object. These results suggest that, just as in humans, a continuum of risk-taking propensities exists in nonhumans.

Let's switch species. Since I was in graduate school, I have been fascinated—"obsessed" might be a better word—with what is called predator inspection behavior in the guppy (*Poecilia reticulata*). Predator inspection is akin to military guard duty: One or a few fish break away from the safety of their group and slowly approach a potential predator to gather information about its possible danger. At first, my interest in predator inspection focused on the cooperative nature of the ways that pairs of fish inspected—ways that matched predictions from economic models of cooperation. But then I realized that predator inspection would also allow me to measure the costs and benefits of risk-taking. After all, I was finding that some fish were consistently willing to take the risk of inspecting predators, while other guppies consistently avoided such risky behaviors. Why? What were the underlying costs and benefits that drove the evolution of risk-taking in this system?

The cost of predator inspection was fairly easy to measure and arguably intuitive. If you take risks in the presence of a predator, you get eaten more often—and, indeed, controlled experimental work, in which groups with different numbers of risk-takers were monitored, confirmed that risk-takers are more likely to end up in the guts of predators (Dugatkin 1992). Measuring the benefits of predator inspection behavior proved to be more conceptually challenging. What were the guppies getting out of these risky endeavors? For natural selection to favor *any* amount of predator inspection behavior, there must be compensating benefits for undertaking this risky behavior, but what are they? My colleague Jean-Guy Godin had a clue when we saw something unexpected. Much research had already demonstrated that female guppies generally prefer more colorful males as mates. What we were seeing was that more colorful males also tended to be risk-takers (Godin and Dugatkin 1996).

We ran a series of experiments designed to see if it was a male's color, his risk-taking tendencies, or both that made him attractive as a potential mate to females. When we experimentally decoupled boldness and color (using a nifty experimental device that a former engineer at NASA built for us), female preference was most directly linked to risk-taking behavior in males. Why females prefer bolder males is still not understood, but it may be that boldness is a signal of genetic quality. What does seem clear is that natural selection has favored males that temper their risk-taking behavior as a function of the benefits available: Colorful males inspect predators far less often when females are not watching them than when females are observing what they do in the presence of a predator.

Attraction to the opposite sex may not be the only good thing to come from risky bouts of predator inspection. My colleague Michael Alfieri and I also found that risk-takers were better at learning associative-memory tasks than were risk-averse fish. We measured a guppy's tendency to inspect predators and, in a separate test, its ability to pair food with an arbitrary cue. Bolder guppies learned this associative-learning task more quickly than did their shyer group mates (Dugatkin and Alfieri 2003). But there is an interesting catch here: Bold fish were better learners only after they had recently expressed their boldness. Why this is so is an area ripe for future exploration.

Birds of a Feather

Over the last two decades, Peter Drent and his colleagues have been studying risk-taking in the great tit bird (*Parus major*). Two overarching personality types emerge in this species. "Fast" birds explore new environments with rapid-fire speed and are quite aggressive. They are the risk-takers. "Slow" birds show the opposite behavior pattern. In a clever experiment involving "tutors," the researchers also found that fast birds would switch habitats when a so-called tutor passed on information that a new habitat was profitable, but slow birds, which spent more time focused on their environment and less on the tutors, paid less attention to tutors (Marchetti and Drent 2000).

Drent and his team followed birds in natural populations with an eye toward studying the mating patterns of fast and slow birds. They found that fast birds that mate with fast birds and slow birds that mate with slow birds are the pairings that produce the most offspring (Both et al. 2005). These researchers have even gone so far as to create experimental "lines" of great tit birds, by mating the fastest birds with the fastest and the slowest with the slowest. They then examine how genetics and development interact to shape personality (Naguib et al. 2011). This work, in conjunction with other studies, shows a clear genetic basis to risk-taking personality traits in this species.

Including the orangutans at the opening of this piece, we have looked at personality in mammals, fish, and birds. But some researchers have argued, and rightly so, that this is a very vertebrate-centered view of personality, and that invertebrates may also have much to teach us about the evolution of risk-taking and personality. Work in this area is scant, but studies on red octopuses (*Octopus rubescens*) and dumpling squid (*Euprymna tasmanica*) have uncovered evidence that they, too, show a bold–shy continuum (Mather 2008; Sinn et al. 2010). One especially appealing aspect of future work in invertebrates is that the nervous system of cephalopods, like octopuses and squid, is both well understood and easy to study (cephalopods are famous in neurobiology for their giant neurons), thus allowing potential insight into the underlying neurophysiology of risk taking.

Applications Near and Far

The study of risk-taking behavior has practical, as well as conceptual, value. For example, conservation biologists have had some success at reintroducing large carnivores, like wolves, into their native habitats. Though these programs should be applauded, they sometimes have unexpected negative consequences, such as the rekindling of old rivalries between the conservation biologists implementing these programs and ranchers who keep domesticated animals on which the reintroduced large carnivores feed. Research on animal personalities might help reduce such tensions.

More specifically, there is mounting evidence that much of the hunting of domesticated animals by wild carnivores is done by a small number of "problem individuals"—that is, the same individuals repeatedly attack and kill ranchers' livestock. Studies of hunting behavior in wolves, cougars, leopards, seals, lions, tigers, bears, and other species all hint at the presence of such "repeat offenders." Attacking ranchers' livestock carries high risk: A predator must first circumvent any fencing or other defensive measures put into place and then risk being shot by ranchers. Predators that consistently attempt to attack such livestock display many of the personality traits associated with boldness (Dugatkin 2009).

Suppose we stop thinking in terms of "repeat offenders" and instead focus on measurable characteristics like boldness. We might examine whether bold predators use particular paths to reach prey, hunt at particular times, or are more or less attracted to certain stimuli than risk-averse predators are. Then, rather than shooting "problem individuals," perhaps wildlife workers could build traps tailored to the hunting strategy of bold predators. Understanding the science of risk-taking might help spare the lives of these carnivores, while at the same time help ranchers protect their stock.

Another practical application of work on animal personalities centers on the use of animals to aid disabled people. To become a guide for blind people, for example, dogs must pass through a series of tests, and the factor that eliminates the most dogs from the pool of possible guides is fear. Some dogs are just more frightened by novel things in their environment, and they are less willing to take risks to help the person they are guiding. Based on this phenomenon, researchers have developed a scoring system for fear and fearlessness and have implemented it across dog breeds to find the breed most likely to aid blind people.

We've made some progress, but much work remains to be done on the evolution of boldness in animals. We need a better understanding of the molecular genetics and the underlying neurobiology and endocrinology of this trait. We need mathematical models of risk-taking; these models are only just emerging in the growing animal literature on behavioral syndromes. We need a deeper understanding of the phylogenetic—that is, evolutionary—history of boldness. Almost nothing is known about this, though

Alexander Weiss has some fascinating ideas about the relationship between the evolution of personality traits in humans and that of our closest living relatives (Weiss et al. 2011).

I'm optimistic on all fronts. The more species we investigate, the more costs and benefits we measure, the more theory we develop, and the more new tools we develop to study the genetics, neurobiology, and hormonal underpinnings of boldness, the better we will understand boldness in non-humans and, eventually, humans.

2

Hit Parade

The Future of the Sports Concussion Crisis

By Chris Nowinski

Chris Nowinski is a co-director of the Boston University Center for the Study of Traumatic Encephalopathy and the co-founder and executive director of the Sports Legacy Institute, a nonprofit organization dedicated to solving the sports concussion crisis. A former Harvard football player and professional wrestler, he is the author of *Head Games*, which was made into a 2012 documentary film. Nowinski was named a 2011 Eisenhower Fellow and serves on the NFL Players Association Mackey-White Traumatic Brain Injury Committee as well as the Ivy League Multi-Sport Concussion Committee.

While concussions have long been linked to brain and central nervous system issues, a new study suggests that repeated hits to the head—mild or otherwise—can lead to memory loss, depression, and dementia. This postmortem brain study, conducted at the Boston University Center for the Study of Traumatic Encephalopathy, provides new and troubling evidence about chronic traumatic encephalopathy (CTE), a long-term degenerative and incurable brain disease. Although military personnel and others are vulnerable to the disease, the highest risk is among athletes involved in contact sports in which hits to the head are considered "part of the game."

⟨⸻⟩

TEN YEARS AGO, few would have predicted that brain injuries would one day dominate the sports headlines. When former NFL star Junior Seau committed suicide in August 2013, the media focused almost entirely on whether the thousands of head blows he endured during his 19-year career as a middle linebacker were a contributing factor. More than 3,000 former NFL players are suing the league for allegedly misleading them about the risks of brain injury, and new policies and studies aimed at protecting the brains of athletes seem to be announced every week. But it's not just professional athletes who are the focus of attention. No fewer than 40 states have passed laws requiring athletes in schools and recreational programs to schedule a doctor's appointment when a concussion is suspected.

I might be a different person today if I had been more aware of the risk I was facing as a football player at Harvard University and, later, as a professional wrestler. In 2003, I suffered a concussion during an in-ring wrestling accident. Despite suffering temporary amnesia, I finished the match. Over the next few days, I developed a throbbing headache every time my heart rate became elevated. Yet I continued to wrestle or exercise nearly every day for five weeks until developing rapid-eye movement (REM) behavior disorder, a condition that triggers abnormal physical movements during sleep.

During my five-year recovery from the concussion, I delved into brain trauma literature and decided that it wasn't worth risking what might develop if I took more blows to the head. The literature, which included article after article pointing to the serious consequences of even seemingly mild brain injury, moved me to dedicate my life to making others aware of the dangers and to work toward developing a treatment for chronic traumatic encephalopathy (CTE). A progressive, degenerative brain disease, CTE can manifest in athletes and others with a history of repetitive brain trauma months, years, or even decades after injury. Memory loss, confusion, depression, aggression, impaired judgment or impulse control, and, eventually, progressive dementia may result. With all the head hits I took, I know I am at risk for the disease. But even if I never suffer from CTE, I suspect that many of my friends and former teammates will.

With this newfound awareness about the dangers of concussion, parents face tough choices about which sports their children should be allowed to play. Some of the more dangerous sports for the brain, such as football, soccer, ice hockey, and lacrosse, are also the most popular. Although everyone agrees that brain trauma may have lasting and debilitating effects, and science continues to make slow progress toward understanding the disease, we cannot yet entirely quantify those effects. As a result, parents and even medical professionals are left to search their hearts and scour websites for answers. But a decade's worth of research has made one thing clear: We need to find better ways to protect the brains of athletes.

Old and New Science

Multiple brain banks throughout North America have recently reported that CTE is commonly diagnosed postmortem in athletes with a history of repetitive brain trauma. And while regular media reports on CTE are a new phenomenon, the problem itself is an old one. It was first described in 1928 as "punch drunk" by a New Jersey medical examiner writing in the *Journal of the American Medical Association,*[1] and pathological case reports, mostly in boxers, have been published for decades. By 1954, the condition was so well established in the public consciousness that Marlon Brando, in the classic film *On the Waterfront*, would win a best actor Oscar for portraying Terry Malloy, a punch-drunk ex-boxer with slurred speech, a poor memory, and a hot temper.

Because the risk factors for CTE have yet to be formally investigated, there is limited understanding of the behaviors that cause or contribute to the disease. We are hyperfocused on concussions—brain injuries that cause clinical symptoms—but there is a poor correlation between number of concussions and the risk and severity of CTE. Most experts once believed subconcussive impacts, in which there are no clinical symptoms, were harmless. Yet new research has found that many athletes who suffer hundreds of impacts during a season, even without a concussion, show the same brain abnormalities that concussed athletes exhibit on imaging tests and multiple measures of brain activity.[2, 3, 4]

These findings have led experts such as Dr. Robert Cantu, medical director of the Sports Legacy Institute in Boston, to suggest that CTE appears most correlated with total lifetime brain trauma, which could be defined as a combination of subconcussive brain trauma and concussions. Many postmortem diagnoses of CTE, in fact, are made in athletes who never had a diagnosed concussion. Nearly all, however, had long careers involving thousands of blows to the head.

Whatever the causal relationship between concussion and CTE, there is no question that CTE has been the driving force behind the new national awareness of concussion. We've known concussions are bad for the brain even longer than we've known about CTE, but what we *did* know about concussion was never enough, on its own, to inspire us to make sports safe for the brain. It is imperative, now, that we do.

Difficult to Measure

Concussions suffer from a perception problem. On the surface, they might not seem to have a lasting, serious impact. They are an invisible injury: There is no blood, there are no displaced bones, and the patient rarely complains. Even when an athlete is knocked unconscious and observers react with panic, the concern quickly fades. Ninety-nine percent of concussed athletes wake up in seconds or minutes and then seem fine. When symptoms persist beyond the day of injury, in the vast majority of cases they dissipate within a month. The injury seems as if it is gone forever, leaving no scars or overt indication that it ever happened.

Even acute negative outcomes are easily missed. Persistent postconcussion syndrome, for example, is rarely witnessed because sufferers are often so impaired that they stop going to school and work or no longer socialize. The worst acute outcome, a condition called second-impact syndrome (SIS), is so rare that only a few cases worldwide are identified each year, meaning few neurologists have ever seen it. The likely cause of SIS is a blow to the head before a prior concussion has resolved. The patient is unconscious within a minute due to rapid brain swelling linked to a loss of autoregulation of the brain, making SIS especially serious. Despite having 50 percent mortality and 50 percent morbidity rates, SIS occurs so rarely that few medical professionals appreciate the risk and, until recently, did not caution against returning concussed athletes into the same game.

In contrast, CTE is not an invisible injury. Postmortem histological staining creates dramatic pictures of diseased and damaged brain cells. Using advanced techniques, Dr. Ann McKee's lab at the Boston University School of Medicine has produced unique images of coronal slices of the two hemispheres of postmortem brains, allowing researchers to observe the diffusion of hyperphosphorylated tau protein. Stained dark brown, these aggregates of tau protein are biomarkers of damaged brain cells. CTE, therefore, can be called a tauopathy. Recent breakthroughs in tau imaging should allow CTE to soon be diagnosed in living people, but for now diagnosis requires postmortem analysis of the brain.

While imaging tau remains in the future, symptoms of advanced CTE have been visible for a long time as they have reduced athletes—including the famous and powerful—into unrecognizable versions of their former selves. Helping our team at Boston University's Center for the Study of Traumatic Encephalopathy better understand CTE was the late Dave Duerson, one of the great NFL strong safeties, a two-time Super Bowl champion, and the 1987 NFL Man of the Year. An honors graduate of the University of Notre Dame, he went on to serve on its board of trustees. He was both a dedicated family man and a successful businessman. By the age of 45, however, he began to suffer headaches, developed cognitive problems, and ended up making poor business decisions that put

The Stages of CTE

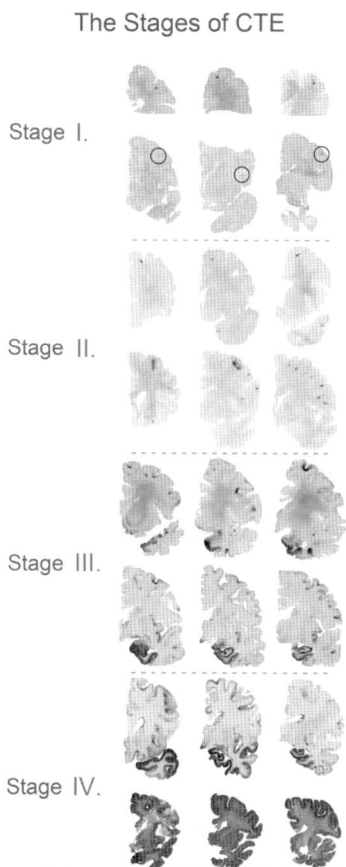

Stage I.

Stage II.

Stage III.

Stage IV.

Figure 1. *The four stages of chronic traumatic encephalopathy. In stage I CTE, p-tau pathology is found in isolated foci in the cortex. In stage II CTE, there are multiple focal areas of p-tau pathology and involvement of nearby brain. In stage III, p-tau pathology is spread throughout the brain including the medial temporal lobes. In stage IV CTE, most regions of the brain show severe p-tau pathology.*

his firm into receivership and himself into millions of dollars in debt. He developed impulse control problems that led to violent outbursts toward his family, leading to divorce. When he took his life in 2011, he left a note requesting that his brain be donated for study by our team.

Our studies revealed that Duerson had sustained damage to his frontal lobes and hippocampus. This damage may have contributed to the cognitive impairment that led to his trouble with such executive functions as planning and organizing. He also had damage to his amygdala, which may have contributed to his emotional outbursts. We still haven't found a way to diagnose CTE in living individuals, nor have we found a treatment. What we know is that CTE tauopathy is associated with memory, cognition, mood, and behavior disorders. Although a direct causative relationship between brain trauma and CTE has yet to be established, everyone in the published literature diagnosed postmortem with CTE had received significant brain trauma, and currently there are no other identified common variables.

Our team recently published findings from a study of the first 85 brains donated to our brain bank by the families of diseased athletes and military veterans.[5] Sixty-eight were found positive for CTE, which lead author McKee broke down into a new four-stage classification system that elucidates the progressive nature of the disease. Of major concern is that in our sample, 34 of the 35 professional football players were positive for CTE, nine of nine college football players, and four of four professional hockey players. Admittedly, a brain bank case series gives no indication of disease prevalence; the population is biased, especially because families are most likely to donate if the donor was exhibiting abnormal symptoms. Nonetheless, the results are disconcerting, and should inspire us to not allow the next generation of athletes to suffer the same fate.

Children at Risk

Until I became involved in CTE research, I never considered that most brain trauma in the industrialized world occurs in children playing sports. Since participation is voluntary, and the rules of recreational sports are malleable, it seems reasonable to make every effort to reform each individual sport, with the goal of reducing risk of concussions and CTE. As logical as that sounds, adoption of brain trauma limits and other protections for athletes has been remarkably slow. Based on what we know today, there are a number of steps we can take to lower the risk of concussion and CTE.

The rules of sports are not static. Committees meet each year to rewrite rules based on new information. Most of these changes have to do with keeping the game fun and entertaining; for example, adding the three-point line in basketball or tweaking the offside rule in soccer. Occasionally, the rules are changed for safety, and no issue has come along that is more crucial than protecting a young athlete's brain function.

Historically, athletes have participated in sports with rules that ignore the risks. Until recently, ice hockey players were allowed to intentionally use their skating momentum to slam into any part of their opponent, including the head, with little concern for penalties, fines, or suspensions.

When I played football, which wasn't that long ago, coaches taught us to lead with our heads as the point of initiating contact for blocking or tackling. Athletes were encouraged to play through concussions if they were able.

New rules in both these sports have since been designed to lessen brain trauma; but with every new horror story that emerges on the sports pages, parents worry even more. What sports should I allow my child to play? What power do I have to protect my child on the field? To evaluate the risk, simply compare how that sport is played at the youth versus adult level, and consider the safeguards professionals are provided. Football is a prime example since, amazingly, 6-year-olds play by essentially the same rules as professionals. Right now we have a healthy national discussion about whether the NFL is too dangerous for adults, yet we pay less attention to the risks of youth leagues, despite the fact that football is far more dangerous for kids.

We need to consider the way the human brain develops and recognize that children are at an anatomical disadvantage compared with NFL players. A child's axons, which connect brain cells to one another, are not fully myelinated (in other words, insulated), and his or her brain cells are more sensitive to the neuron-damaging shock of concussions, making each impact and concussion potentially more damaging to the brain.[6]

Children are also at a biomechanical disadvantage. A child's head grows much faster than his or her body, so the head is nearly fully grown by the age of 4, a time when body mass is about 20 percent of full size. Even by age 12, when a child's head is 95 percent of its eventual full size, his or her body is only half its eventual full mass.[7] Combine the child's mature head with a weak neck and torso, and a comparison might be made to a bobble-head doll. It doesn't take much force to accelerate the head to dangerous levels, such that the brain pitches back and forth and twists within the skull, producing chemical, metabolic, and even structural injury to the brain. In fact, studies involving sensors in helmets have revealed that children take blows to the head of almost equal force as do college players.[8]

When it comes to diagnosing concussions, children also face inadequate safeguards. There is no biomechanical or neuroanatomical reason to believe that children aren't suffering as many concussions as adults, and yet they are rarely diagnosed with the injury. A few years ago sports leaders believed that children didn't actually suffer concussions—they were somehow resilient. Now we know there are two major reasons children aren't frequently diagnosed with concussions.

First, rarely is anyone on hand to diagnose the injury and, second, young players seldom report symptoms. Your average NFL team has multiple medical professionals at every game and practice. Your average youth football game or practice has no medical personnel on hand. A recent study found that high schools with athletic trainers diagnosed eight times as many concussions as high schools without medical staff.[9] Another study found that medical doctors who aggressively evaluated hockey players displaying concussion symptoms diagnosed seven times as many concussions as teams that had only athletic trainers on the bench.[10] Do the math: If we provided children with athletic trainers and doctors on the sideline, we'd diagnose about 56 times more concussions. By not providing these resources, a solid case can be made that we will continue to miss 55 of every 56 concussions.

If this statistic seems hard to believe, consider that most concussions are not diagnosed unless the player self-reports symptoms. Educational programs from the Centers for Disease Control and Prevention advise players, "It's better to miss a game than the season"—an effective message for prompting high school and college athletes to consider their long-term futures and self-report their symptoms. Young athletes, however, are not likely to have the cognitive capacity to recognize their symptoms as being connected to trauma, nor to realize they should inform an adult. Moreover, such messaging is rarely provided to children. Because there is no validated educational program for child athletes, parents, coaches, and other adults must actively teach youngsters about concussions and also encourage them to report their own symptoms or those of a teammate.

Recommended Reforms

When reviewing the evidence, as well as the empirical data, it is hard to argue that football isn't dramatically more dangerous for children than it is for adults. Even simple protective measures, such as reducing full contact in practices, are more common in professional than youth leagues. NFL players average full-contact practices less than one day per week during the season, while the vast majority of youth programs set no limits. Youth football needs to follow the lead of other sports that have recognized the difference between children and adults and have instituted aggressive rule changes that reduce the risk of concussion and repetitive brain trauma. USA Hockey has increased the age at which checking—a source of many concussions—is allowed, from 11 to 13. Some soccer organizations recommend not introducing heading, a basic fundamental of the sport, until at least age 10.

To further reduce exposure to brain trauma, the Sports Legacy Institute has launched the Hit Count Initiative, which is modeled on youth baseball's Pitch Count System. This system was developed after data revealed that the more times a pitcher threw the ball in a day, week, or year, the greater the risk of wear and tear to the ulnar collateral ligament. Although elbow injuries are rare in youth baseball, every player is now limited in the number of times he or she can throw in a day and is required to take multiple days of rest after pitching exposure. Dr. Cantu and I believe that such exposure limits make even more sense for the brain. In high school football, for example, a player takes nearly 700 blows to the head each season that have forces of 10g or greater, and some players have recorded almost 2,500 hits.[11] More than half of these hits are sustained in practice, meaning they can be eliminated without dramatic change to the game.

Sensors in helmets, headbands, skullcaps, mouthpieces, and chinstraps will soon become widely available and inexpensive. The Sports Legacy Institute is working with several manufacturers of these products to define the threshold for a hit and to develop recommendations for exposure limits. College football programs at Virginia Tech, Dartmouth, Brown, North Carolina, and Oklahoma have started using the Head

Impact Telemetry System, which costs around $50,000 to $60,000. For high school and college programs, this system can make a real difference, especially if the school's authorities choose to tackle the tricky question of how many severe hits are too many for a player. I envision a day when the price tag comes down for measuring hits to the head, allowing a real-time count to be accessible to all parents and coaches. Armed with that knowledge, coaches can adjust practice habits to limit exposure, and coaches can identify players who are utilizing techniques that may be putting their brain at greater risk. I also hope that parents will be able to use this information to inspire coaches to effect change quickly.

Technology can make us safer, but it's only one piece of the puzzle. There is no simple answer. Every sport involves risk, just as crossing the street does. While we can quantify the risk of driving in a car or jumping from a plane, without a way to diagnose CTE in living individuals, we can't quantify the relative risk of each sport. We are left with case studies, media stories, hypotheses, and confusion.

If we were starting from scratch, armed with our current knowledge about sports-related brain injury, I am confident we would limit children's participation to those sports in which head trauma is both *rare* and *always accidental*. Perhaps that's a guideline for parents to consider. We must remember that many of the world's most popular — and entertaining — sports were designed for fully mature adults who are competing with informed consent. Professional sports are designed to sell tickets and increase television ratings. Children's sports should focus less on viewership and more on improving health, teaching lessons, and building character. More troubling is the future of high school and college football, where aspects of both worlds collide. The issue has reached a fever pitch with more and more professional football players expressing a fear for their future health and the future of their sport every day. Even President Barack Obama recently entered the conversation, expressing concern about college football players and the NCAA's role in addressing "problems with concussions and so forth." The "so forth" he alludes to clearly is CTE.

There is little doubt that athletic activity is great for children, but now there is also little doubt that science is revealing the grave consequences of repetitive brain trauma. In the future, our children should play sports with rules that are designed around the limits of their brains, rather than the limits of their will.

3

Epilepsy's Big Fat Answer

By John M. Freeman, M.D.

John M. Freeman, M.D., was the Lederer Professor Emeritus of Pediatric Epilepsy and professor of neurology and pediatrics at the Johns Hopkins Berman Institute of Bioethics. A graduate of Amherst College and the Johns Hopkins School of Medicine, he initiated the Division of Pediatric Neurology at Johns Hopkins in 1969 and served as its director for 23 years. He also formerly directed the Division of Pediatric Epilepsy and the Birth Defects Clinic at Johns Hopkins. Freeman is the co-author of *Seizures and Epilepsy in Childhood: A Guide for Parents* and *Tough Decisions: A Casebook in Bioethics.*

IN MEMORIAM

The author of the article, John Freeman, passed away on January 3, 2014. Two Dana Alliance members, Guy McKhann, M.D., and Donald L. Price, M.D., both professor emeriti in the Departments of Pathology, Neurology, and Neuroscience at Johns Hopkins University School of Medicine, remembered him:

"John's resurrection of the ketogenic diet, which completely eliminated the epileptic seizures of many patients, was accomplished virtually all by himself, against great skepticism and opposition." —Guy McKhann

"John was one of the original cohorts of neurologists involved in the creation of the Department of Neurology at Johns Hopkins. He was a thoughtful and compassionate clinician committed to pediatric patients and their parents. Moreover, he was a world expert in epilepsy. We were friends/colleagues for more than 40 years and all of us loved his sense of humor. John was a real family guy and a superb person. He will be greatly missed." —Donald L. Price

---

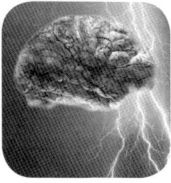

Epilepsy and seizures affect nearly 3 million Americans of all ages. The incidence is greater in African-Americans and in socially disadvantaged populations, and about 200,000 new cases of epilepsy are diagnosed each year. Despite these alarming figures, no magic pill exists to eliminate convulsions. While drugs work for some, others find them ineffective. What seems to work just as well, if not better, especially in children, is a relatively unknown, high-fat diet. The author, John M. Freeman, M.D., one of the nation's leading advocates for its use, writes about the evolution of the diet and its struggle for acceptance.

It took a Hollywood producer, a made-for-TV movie with Meryl Streep, two decades of research, and the retraining of a cadre of dietitians and physicians to bring the ketogenic diet back into the mainstream. One of the first and most successful therapies for epilepsy, the diet had fallen into obscurity with the introduction of anticonvulsant medications in the 1930s. Today, however, not only has the diet reemerged as a therapeutic option in epilepsy, but its effectiveness for other neurological disorders, including brain tumors, autism, and even Alzheimer's disease, is being explored as well.

The hallmark feature of the diet is the production of ketones (beta-hydroxybutyrate, or BOHB), the residues left when fats are burned in the absence of sufficient glucose. Glucose is an important source of the brain's energy, but, contrary to popular belief, it is not the only potential source. Ketones are, in fact, a more efficient energy source for the brain and, for unknown reasons, make the ketogenic diet—high in fat, low in carbohydrates—more effective than current anticonvulsant medications in curbing difficult-to-control seizures.[1,2]

Our hunter-gatherer ancestors, who underwent long periods of quasi-starvation as they searched for game, burned their body fat, producing ketosis, and used BOHB to fuel their brains. Ancient Greek physicians treated diseases, including epilepsy, by altering their patients' diets. But it wasn't until the early 20th century that modern physicians stumbled upon clues that led them to use this tactic purposely, even if they didn't truly understand it.

During the pre-insulin era of the 1920s, starvation became a treatment of choice for managing children with diabetes, who lacked insulin to transport glucose to the brain. The morbid effects of diabetic ketoacidosis (lethargy, coma, and even death), however, led to the misbelief that ketosis per se was dangerous. In 1922, Hugh Conklin, an osteopath and faith healer from Battle Creek, Michigan, erroneously theorized that epilepsy arose from the intestine and attempted a cure by fasting a number of children with epilepsy for up to 25 days, providing only limited liquids. This starvation, he found, controlled seizures, and its beneficial effect was reported by Rawle Geyelin at the American Medical Association convention in 1921.[1] Soon after, Russell Wilder of the Mayo Clinic reported that the effects of starvation could be mimicked by a diet high in fat and low in carbohydrates and documented that it was as effective as starvation for the treatment of epilepsy in children. Because this diet caused ketosis, it became known as the ketogenic diet.

Following Conklin's and Wilder's lead, physicians often recommended the ketogenic diet, resulting in seizure control in 30 to 50 percent of children placed on it. The diet was far more effective than phenobarbital and bromides, then the only available anticonvulsant medications. With the discovery of phenytoin (Dilantin) in 1938 and the introduction of other anticonvulsant drugs soon afterward, the diet was gradually abandoned and largely forgotten. Yet, even today, with many new medications, 30 percent of children with epilepsy continue to have difficult-to-control seizures.

Then, as now, physicians found it easier to prescribe a pill than to teach their patients all that was required for the preparation of food within a rigid diet—and patients found it easier, too. Consequently, few physicians recommended the diet and not many dietitians knew of its benefits or were trained in its appropriate management.

In 1968, Guy McKhann, M.D., director of child neurology at Stanford, was asked to direct the newly created neurology department at the Johns Hopkins Medical Institutions. I was working under him at the time, and he invited me to join him to create and direct its new pediatric neurology section. Samuel Livingston, M.D., also at Johns Hopkins, directed one of the country's few centers for childhood epilepsy. There the diet remained in vogue and was administered under the supervision of Millicent Kelly, his dietitian. When Livingston retired in 1973, I was asked to become director of the seizure clinic as well, since I was familiar with the management of seizures. In 1990, I ceased directing the pediatric neurology division and focused my attention on pediatric epilepsy.

Although officially retired, Kelly continued to help me and my colleagues use the diet to treat six to eight children each year successfully. One of the few dietitians familiar with management of the diet, she designed specific meals—liquid, soft, Kosher, etc.—which helped parents whose children were on the diet. That meant prescribing foods that had twice the fat content of a McDonald's Happy Meal; it also meant counseling patients to drink heavy cream and eat butter without bread. Without Kelly's knowledge and 40-plus years of experience, we would not have been able to prescribe the diet. Together, we were the keepers of the flame, and we prepared for the flood of patients who were to come.

Diet vs. Drugs

An epileptic seizure is defined as a transient symptom of abnormal excessive or synchronous firing of some or many of the brain's cells, or neurons. The outward effect can be as dramatic as convulsions with wild thrashing movements (tonic-clonic seizure) or as mild as a brief loss of awareness (absence seizure). Sometimes seizures consist of repeated full body "slumps," with the person simply losing body control and crashing to the ground. Recurrent seizures are termed epilepsy. Which of these manifestations occurs depends on which neurons "fire" synchronously.

Epilepsy is usually controlled, but not cured, with medication. It has been found that initial, thoughtfully chosen medication can make almost 50 percent of patients seizure-free for extended periods of time. If an initial drug fails, another well-chosen drug may make an additional 14 percent of patients seizure-free. If that drug fails, too, then the likelihood of rendering someone with epilepsy seizure-free is poor. More than 30 percent of patients with epilepsy will not have seizure control even with the best available medications. Despite the introduction of many new anticonvulsant medications, these figures have remained consistent over time.

Working with families whose children have epilepsy, we realized that there were no books written for parents to help them cope. To fill the void, my colleague, Eileen Vining, M.D., my coordinator/counselor, Diana Pillas, and I decided to team up to write one. The result was *Seizures and Epilepsy in Childhood: A Guide for Parents,*[3] published by the Johns Hopkins Press in 1993. It contained three pages on the ketogenic diet. Sensing a need for a separate book that focused solely on the diet, the dietitian Millicent Kelly and I collaborated with my daughter, Jennifer Freeman, a freelance writer, to write *The Epilepsy Diet Treatment: An Introduction to the Ketogenic Diet,* a shorter book specifically about the ketogenic diet. Sadly, we could not find a publisher.

A New Foundation

It was 1993, and Charlie Abrahams, the 2-year-old son of Jim Abrahams, the Hollywood producer of *Airplane* and *Naked Gun,* continued to suffer many uncontrollable drop seizures each day, despite extensive medical intervention. As Abrahams has stated, "After thousands of epileptic seizures, an incredible number of drugs, dozens of blood draws, eight hospitalizations, a mountain of EEGs, MRIs, CAT scans, and PET scans, one fruitless brain surgery, five pediatric neurologists in three cities, two homoeopathists, one faith healer, and countless prayers, Charlie's seizures continued unchecked, his development delayed, and he had a prognosis of continued seizures and progressive retardation."

Researching epilepsy treatments himself, Abrahams found our book, *Seizures and Epilepsy in Childhood: A Guide for Parents,*[3] with its three pages on the ketogenic diet. After a phone call, he brought Charlie to Johns Hopkins, where the toddler first fasted, according to our protocol. Within several days his seizures disappeared and he began the diet. Gradually taken off his medications, Charlie has remained seizure-free, on no medications, for the past 19 years.

Jim was outraged that in all his conversations with medical experts and other parents, he had never been told of the diet. Determined to make information about the ketogenic diet available to parents and physicians, he was instrumental in bringing to the public the 1994 "Dateline" news magazine program "An Introduction to the Ketogenic Diet," featuring his friend, the actress Meryl Streep. The same year he created the Charlie Foundation to Help Cure Pediatric Epilepsy. The foundation funded a seven-year study and published 2,500 copies of our shorter book, which sold quickly and attracted a more established publisher, DemosHealth.

The foundation also funded the production of several DVDs explaining the diet to parents, dietitians, and physicians. Despite offering them free to physicians at national and regional epilepsy meetings, there were few takers—a sign that much work still needed to be done. A big breakthrough in promoting the diet finally came in 1997, when an Abrahams-directed, made-for-TV movie, *First Do No Harm,* also with Streep, was followed by a flood of thousands of phone inquiries and about 150 patients seeking help with the diet from Johns Hopkins.

The new patients allowed us to gather important data about the diet's effectiveness and its side effects. The reports from Johns Hopkins were the first of an avalanche of abstracts and articles on the clinical outcomes of children who were treated, including outcomes of the diet's various aspects and modifications. The huge increase in the number of clinical abstracts was presented annually at American Epilepsy Society meetings.

The foundation has since been the moving force behind increasing physicians' knowledge about the diet and in training parents and dietitians in its use. At the Third International Conference on Dietary Therapies for Epilepsy and Other Neurological Disorders, organized by the foundation in 2012, close to 500 physicians, dietitians, and parents from 30 countries around the world honored Streep for her role in reintroducing the diet. The conference—promoted in the foundation's newsletter, *KetoNews*—included cooking demonstrations and exhibits, testimonials from parents, and a professional symposium.

Anecdotal and Empirical Evidence

At first, most epileptologists did not believe that a diet could be as effective as drugs, although multiple large studies documented the diet's effectiveness. But these studies were uncontrolled, and since many were based on the large Johns Hopkins patient population, there was a tendency for physicians to discount the results as biased by enthusiasts. Finally, two blinded crossover studies—one from the Cross group in England[4] in 2008, the other funded by the National Institutes of Health at Johns Hopkins in 2009—documented the diet's effectiveness in children in a controlled fashion. Although there are currently no large studies in adults, anecdotal reports indicate the diet's effectiveness, although it appears adults have more difficulty adhering to its rigidity.[5]

The outcome of children with uncontrolled seizures who are placed on the diet is shown in the figure below, which summarizes data from Johns Hopkins studies[6] and is similar to many reports from other centers. It is notable that 33 percent of the children with intractable seizures were s eizure-free, or had only rare seizures, after being on the diet for one year, and 27 percent of the children whose seizures had previously been uncontrollable by medications had no seizures or only rare seizures three to six years after initiating the diet, although by that time most of the children were off the diet and all medications. Some with continued seizures remain on the diet because it has decreased the number of medications the children need to take as well as the consequent side effects. No anticonvulsant drugs have been studied for that duration or have shown the rate of beneficial effects.

The diet is rarely used as the initial treatment for epilepsy, but should

Outcomes of Children on the Ketogenic Diet 1 year and 3-6 years after initiation
% of original use

Number initiating	Seizure control	1 year	3-6 years later
Total N=150	Seizure-free	7%	13%
	>90%	26%	14%
	50-90%	22%	16%
	<50%	5%	12%
	On Diet	83 (55%)	18 (12%)

be strongly considered when two anticonvulsants, properly used, have failed. However, the diet may be the initial treatment of choice in infantile spasms and other "drop" seizures.

Despite considerable recent research, how the diet exerts its beneficial effects remains unknown.[6] It is not solely the ketosis, the accompanying acidosis, the lipidosis, or any of the other chemical changes that have been investigated that are responsible. Glucose restriction appears to be a partial answer. A recent study[9] suggests that episodic fasting, in addition to the diet, is even more effective than the diet alone in controlling seizures. Learning the mechanisms by which the diet controls epilepsy may give us a better understanding of epilepsy itself.

Other Ketogenic-Like Diets

Modifications of the ketogenic diet, such as the modified Atkins diet, the medium-chain triglyceride (MCT) diet, and the low-glycemic diet (LGD), have been developed as alternatives to the rigidity of the classic version. These diets, all of which have been found to have degrees of effectiveness, may be more acceptable to adolescents and adults.[5] Although large studies of these diets have yet to be performed, uncontrolled studies suggest that they may be effective. If tried, and the individual's seizure control is less than satisfactory, the more rigid, classic ketogenic diet is recommended.[1]

While the classic ketogenic diet may cause complications[1] such as kidney stones,[9] lipidemia, and gastrointestinal symptoms, problems are rarely serious and easily managed. Vomiting is common in the early stages of fasting and may be relieved with small doses of glucose. Constipation is common and may be relieved with small amounts of medium-chain triglyceride (MCT) oil or a readily available laxative such as MiraLAX. Carnitine is rarely needed, but sugar-free multivitamins and minerals such as calcium are recommended. Calcium oxalate and uric acid kidney stones occur in 15 to 20 percent of patients and can be treated or prevented by the administration of potassium citrate. Plasma lipids are also known to rise slightly, but they return to normal levels after six months. The rare patient with familial dyslipidemias has been seen, and therefore lipid levels should be checked occasionally.[1]

Generally, children whose seizures are controlled by the diet are tapered off it after two years. But if seizures continue, or recur, the child should remain on the diet longer. Some patients have remained on the diet for more than 25 years without adverse effects.

Other Possible Uses

The resurgence of interest in the ketogenic diet has led to some very preliminary studies of its use in conditions other than epilepsy. Anecdotal reports are leading to controlled studies in various neurological conditions. Neurodegenerative disorders provide a unique opportunity to study cellular protection through diet.[6,9,11]

Animal models and anecdotal human reports suggest that glucose restriction and the ketogenic diet may have beneficial effects on brain tumors.[7,8] Tumors rapidly metabolize glucose but are unable to utilize ketones as an energy source. Brain tissue, on the other hand, is able to use both. Glioblastoma implanted in rodents rapidly regresses on a glucose-restricted ketogenic-like diet. Anecdotal reports and preliminary studies in humans have found tumor regression on ketogenic diets. The dramatic findings in glioblastoma may also be true of other brain tumors and perhaps other systemic tumors.

There are also anecdotal reports of the diet's benefits[6,9,11] in modifying Alzheimer's disease, Parkinson's, amyotrophic lateral sclerosis, and possibly posttraumatic brain injury, stroke, and severe hyperactivity. No controlled studies of these conditions have yet been reported. However, such studies are needed to prove or disprove the diet's usefulness. The diet, or a modified form of it, may also be useful in the management of diabetes. Preliminary reports of its use in inflammatory disease and the management of pain are also of interest and deserve further study.[6]

Although it has taken 20 years to reintroduce the once-abandoned ketogenic diet, it has become an important new therapy—especially for difficult-to-control seizures in children. As more dietitians are trained, and more physicians become aware of the diet, it is used increasingly through-out the world.[1] Guides to its use are now available in many languages and in many countries, thanks to the Charlie Foundation and its British cousin, Matthew's Friends. The idea of food as medicine has been a controversial topic in this country for many years. But the statistics don't lie, nor do the hundreds of young people who will tell you how their lives were changed because of it.

4

Psychiatric Drug Development
Diagnosing a Crisis

By Steven E. Hyman, M.D.

Steven E. Hyman, M.D., is the director of the Stanley Center for Psychiatric Research at the Broad Institute, the Harvard University Distinguished Service Professor of Stem Cell and Regenerative Biology, and a member of the Dana Foundation Board of Directors. Hyman joined the Broad after a decade of service as provost of Harvard University, where, as Harvard's chief academic officer, he focused on the development of collaborative scientific initiatives. From 1996 to 2001, he served as director of the U.S. National Institute of Mental Health (NIMH). Hyman is the editor of the *Annual Review of Neuroscience* and the founding president of the International Neuroethics Society. Prior to his government service, Hyman was the first faculty director of Harvard University's interdisciplinary Mind, Brain, and Behavior Initiative, where he studied the control of neural gene expression by neurotransmitters with the goal of understanding mechanisms that regulate emotion and motivation in health and illness.

Although one in five Americans currently takes at least one psychiatric drug and mental disorders are recognized worldwide, global pharmaceutical industry funding for new, innovative medications is in serious decline despite promising advances in genetics. The author traces the evolution of psychiatric drug development, the reasons for its retreat, and the changes necessary to meet the growing demand.

During the past three years the global pharmaceutical industry has significantly decreased its investment in new treatments for depression, bipolar disorder, schizophrenia, and other psychiatric disorders.[1] Some large companies, such as GlaxoSmithKline, have closed their psychiatric laboratories entirely. Others, such as Pfizer, have markedly decreased the size of their research programs. Yet others, such as AstraZeneca, have brought their internal research to a close and are experimenting with external collaborations on a smaller scale.

This retreat has occurred despite the fact that mental disorders are not only common worldwide, but also increasingly recognized by health-care systems. There is, moreover, vast unmet medical need, meaning that many individuals with mental disorders remain symptomatic and often disabled despite existing treatments. For example, people suffering with the depressed phase of bipolar disorder often continue to experience severe symptoms even when they take multiple medications with serious side effects. For some significantly disabling conditions, such as the core social deficits of autism and the cognitive impairments of schizophrenia, there simply are no effective treatments. Because mental disorders are highly prevalent and our ability to treat them remains limited, these illnesses cause enormous societal burden. In aggregate, they are the world's leading cause of disability.[2]

In addition, this retreat has happened despite the fact that different classes of psychiatric drugs have been among the industry's most profitable products during the last several decades—and despite the fact that, according to Medco Health Solutions, one in five American adults now takes at least one psychiatric drug. Among the earliest commercial successes were the Valium-like benzodiazepines, used both as tranquilizers and as sleeping pills. These were followed by the Prozac-like selective serotonin reuptake inhibitor (SSRI) antidepressants. Most recently, "second-generation" antipsychotic drugs have been among the global revenue leaders for the pharmaceutical industry, serious side effects notwithstanding. That's why it's surprising that almost all industry research dollars are invested in cancer, metabolism, autoimmunity, and other disease areas. As the expiration of patents on blockbuster drugs squeeze budgets, companies perceive their withdrawal from psychiatry as an unfortunate but rational reallocation of research resources.[3] This withdrawal reflects a widely shared view that the underlying science remains immature and that therapeutic development in psychiatry is simply too difficult and too risky.

The Diagnosis

The scientific issues facing translational psychiatry—the application of basic discoveries in neuroscience, genetics, and psychology to understanding disease and to advancing therapeutics—are daunting. The molecular and cellular underpinnings of psychiatric disorders remain unknown; there is broad disillusionment with the animal models used for decades to predict therapeutic efficacy; psychiatric diagnoses seem arbitrary and lack objective tests; and there are no validated biomarkers with which to judge the success of clinical trials.[4,5] As a result, pharmaceutical companies do not see a feasible path to the discovery and development of novel and effective treatments. Given the steady stream of drugs that have gained approval during recent years for treating depression, anxiety, schizophrenia, and bipolar disorder, this scientific stall may have seemed to come out of the blue. However, payers (both insurance companies and governments) and regulatory agencies have given up their willingness to accept even more expensive new drugs that, despite marketing efforts, have turned out to be no more than variations on very old themes.[3,4]

Even if current drugs recycle old action in the brain, the existing pharmacopeia is a great blessing to many patients and their families. That said, progress for the many patients who respond only partially or not at all to current treatments requires the discovery of medications that act differently in the brain than the limited drugs that we now possess. The molecular actions of all widely used antidepressants, antianxiety drugs, and antipsychotic drugs are relatively unchanged from their 1950s prototypes. Current antidepressants alter levels of the neurotransmitters serotonin or norepinephrine in synaptic connections between certain nerve cells in the brain. This is the same basic action of the first modern antidepressant imipramine, discovered in 1957. Antipsychotic drugs act on several different neurotransmitter receptors in the brain, but the critical shared mechanism of all current antipsychotic drugs is blockade dopamine D2 receptors, the same mechanism of the prototype antipsychotic drug chlorpromazine, discovered in 1950.

More problematic is the failure to improve efficacy, although significant progress has been made since the 1950s on safety and tolerability. Even the most recent antidepressants are no more effective than imipramine. The early antidepressants could prove deadly in overdose; selective SSRIs and other modern antidepressants are far safer and have far milder side effects, permitting far wider use. All antipsychotic drugs, with the exception of clozapine (discovered in the 1960s), have roughly the same efficacy as chlorpromazine. Clozapine clearly benefits some patients with schizophrenia and bipolar disorder, even when other drugs have failed. But the basis of its greater efficacy remains mysterious. Moreover, clozapine's very severe side effects limit its use.

Notably, existing antipsychotic drugs, including clozapine, treat only a subset of the symptoms of schizophrenia, such as hallucinations and delusions. None of the existing drugs improve the cognitive schizophrenia symptoms that are responsible for much disability. The second-generation antipsychotic drugs have far less tendency than older drugs to cause serious motor system side effects, some of which mimic Parkinson's disease. But the newer drugs carry their own burden of serious side effects, such as significant weight gain and elevated levels of glucose and lipids.

Arriving at the Crossroads

The mid-20th century saw the birth of modern psychopharmacology, specifically the discovery of the first medicines that could effectively treat symptoms of specific disorders. Serendipitous observation followed by intelligent follow-up played an important role in the history of medicine, famously including Alexander Fleming's discovery of penicillin. Fleming studied rather than discarded the mold-contaminated petri dishes on which he had plated bacteria. Similar serendipity was involved in discovering the utility of lithium as well as the prototype antipsychotic, antidepressant, and benzodiazepine drugs.

In 1949, John Cade, who was interested in the properties of uric acid, recognized that it was the lithium moiety of his lithium urate salts that was sedating his guinea pigs. This led him with breathtaking rapidity to test lithium on patients with mania. In 1950, the French surgeon Henri Laborit tested the new drug chlorpromazine, originally developed as an antihistamine, as a medication to be used before general anesthesia. Based on chlorpromazine's sedating properties, he recommended that his psychiatric colleagues, Jean Delay and Pierre Deniker, test it on agitated psychotic patients. Remarkably, the sedation turned out to be a side effect; the true benefit of chlorpromazine (later branded as Thorazine) was its ability to diminish the hallucinations and delusions of patients with schizophrenia and related disorders.

As chemists attempted to improve upon the three-ring structure of chlorpromazine, one of the compounds that emerged, imipramine, failed to treat psychosis but markedly elevated mood. Imipramine was then developed as the first of the tricyclic antidepressants, and it became the prototype antidepressant that increases the concentration of the monoamine neurotransmitters (serotonin or norepinephrine) in synapses. It works by blocking the reuptake "pump" that normally removes these neurotransmitters from synapses after they have delivered their signal. The other major antidepressant mechanism, monoamine oxidase inhibition, was based on yet another serendipitously identified drug, iproniazid. It was synthesized in attempts to produce chemical alternatives to isoniazid, a drug used to treat tuberculosis. Iproniazid failed to treat tuberculosis but markedly improved the depressed mood of the chronically ill patients in its clinical trial.

Pharmaceutical companies wanted to follow up on these remarkable discoveries in the 1950s in order to produce additional revenue-generating treatments. Today, in most fields of medicine, scientists are able to identify the molecules in the brain with which these drugs interacted and then exploit these drug targets to develop new medications. Scientists might identify new targets unrelated to existing drugs via knowledge of the genetics of disease risk or understanding of the molecular mechanisms of the illness. In the 1950s and 1960s, however, the biochemical and molecular tools for identifying neurotransmitter receptors did not exist, and the existence of neurotransmitter reuptake transporters was not known. (The discovery of antidepressant and antipsychotic drugs motivated research in the 1950s and 1960s that would eventually yield Nobel Prizes for Julius Axelrod in 1970 and Arvid Carlsson in 2000. Axelrod and colleagues discovered neurotransmitter reuptake mechanisms; Carlsson recognized that antipsychotic drugs must work by blocking dopamine receptors.)

Lacking molecular tools, pharmacologists working in psychiatry developed assays, mostly in laboratory rats, based on the effects of prototype drugs such as chlorpromazine, imipramine, and chlordiazepoxide (Librium) on animal behavior. For example, a rat placed in a beaker of cold water will swim for a time, but it will eventually stop swimming and begin to float. The antidepressant drug imipramine prolongs the period during which the rat attempts to swim. This forced swim assay was rationalized with anthropomorphic terms such as "behavioral despair," but it was never shown to model human depression. Using the forced swim and a battery of other assays, new compounds were screened for possible antidepressant efficacy. Compounds that acted like imipramine were deemed candidate antidepressants, as long as they were not too toxic, and often tested in human clinical trials. Such assays ended up being black box drug screens.

Despite some degree of surface plausibility—the forced swim result seemed to mimic learned helplessness, a putative model of depression—the mechanism by which imipramine causes continued swimming is still not known. Moreover, imipramine increases the duration of swimming with a single dose, whereas depressed human beings generally require several weeks of treatment before therapeutic effects emerge. A large number of compounds that passed such black box assays were ultimately approved as antidepressant, antipsychotic, and anxiolytic drugs; thus, the approach seemed to be succeeding. From the very beginning, however, astute scientists predicted that this approach to drug development would result in the identification of only "me too" drugs. Unfortunately, during the last 50 years that concern has been fully borne out. Perhaps the most troubling aspect of excessive reliance on these black box assays is that potentially efficacious drugs with novel mechanisms of action (e.g., possible antidepressants that are different in mechanism from imipramine) may well have been screened out and discarded.

Since the 1950s, scientists have worked to go beyond these assays and to create animal models of human depression, bipolar disorder, schizophrenia, and other psychiatric disorders. Relying mostly on laboratory rodents, they have used a range of tools from environmental stressors to the insertion of human disease risk genes into the brains or germ lines of mice. If we are to be clear-eyed about the results to date, we would have to conclude that none of these models have proven adequate. Animal research remains critical for basic science, including investigations of basic molecular pathways on which drugs act. However, the use of animals to produce "good enough" models of real human diseases requires that the disease mechanisms be conserved in evolution between the animal selected and human beings. In short, psychiatry will advance not by rejecting animal research, but by showing appropriate circumspection about the use of animals to model diseases.

Another important lesson is that even effective drugs may not prove to be useful keys to understanding disease mechanisms. Even if drugs that block dopamine receptors treat psychotic symptoms, it does not follow that the fundamental problem is excess dopamine any more than pain relief in response to morphine suggests that the original problem is a deficiency of endogenous opiates. If the gains made in other fields of medicine are to serve as models for psychiatry, it is time to make new attempts to understand fundamental disease mechanisms and to apply what is learned to therapeutics.

Discovery and Development

Industry plays a critical role in producing new treatments. There is overlap between research in industry and in academia—each realm has different core strengths. Academics are able to tackle projects that are too risky and long-term for companies focused on increasing stock prices or returning dividends to shareholders. The equation also includes the funders of academic science—largely governments and foundations, which also have a far longer time horizon than do investors in private companies. Funded by tax and philanthropic dollars, academics can accept greater risk and can recognize that the significant leaps into the unknown by which science progresses may have no obvious practical implications in the short run, and that creative explorations commonly produce failure more often than they produce success.

Academia is therefore in a better position than industry to investigate basic mechanisms of disease, as well as other matters that may ultimately be relevant to therapeutics but are still distant from the design of products. Such research can reveal genes, proteins, or other molecules that, if activated, blocked, or otherwise modified, might exert therapeutic benefits and be labeled drug targets. If the research convincingly demonstrates a potential role in the disease processes, the research may earn a "validated drug targets" label. At this point, pharmaceutical companies generally need to take the next step, which is to search for chemical compounds that bind to and modify the target in desired ways. It is the role of medicinal chemists, most often company based, to painstakingly synthesize and study variations on promising chemical compounds in the search for a good drug—a chemical compound that has the desired effect on the target, that is not too toxic, that can be absorbed into the body, and, in the case of psychiatric medicines, that can enter the brain. Once identified, the promising drug must be tested in humans and shown to have the desired therapeutic effects without disproportionate side effects. The clinical trials that take a chemical compound from the stage of discovery to approval by regulators such as the Food and Drug Administration are extensive and costly. Large companies have the necessary resources and experience to orchestrate such trials, which are critical steps in producing new medicines for psychiatric disorders.

Recently, venture capitalists have been exploring ways to pick up the slack, but so far they have not made a significant difference. While they might make an initial investment in a biotechnology start-up, it is far more likely that their return will come through an initial public offering of stock (assuming that the start-up is able to create a successful product) and/or a decision to sell off the start-up to a large pharmaceutical company. As large companies withdraw from psychiatric drug development, interested parties have considered various solutions, including public-private partnerships that would involve government, academia, and industry in an attempt to decrease the risks inherent in drug discovery and development. One such partnership, the Alzheimer's Disease Neuroimaging Initiative (ADNI), has found potentially useful biomarkers for clinical trials that could be used by any company.[6] Such partnerships may help, but since knowledge concerning most psychiatric disorders is less advanced than that concerning Alzheimer's disease, it is far more important to encourage basic understandings of disease processes. In short, without more basic advances, the climate doesn't seem favorable yet for the creations of partnerships to find biomarkers for psychiatric disorders.

Recent Advances

As long as we guard against renewed self-deception about what constitutes meaningful advances, there is good reason to feel optimistic about the long-term future of translational psychiatry—despite its palpable scientific challenges. My optimism is based partly on the extraordinary vitality of neuroscience and perhaps, even more important, on the emergence of remarkable new tools and technologies to identify the genetic risk factors for psychiatric disorders, to investigate the circuitry of the human brain, and to replace current animal models that have failed to predict efficacious new drugs that act by novel mechanisms in the brain. New ideas are, of course, central to scientific progress, but new tools can open up unexpected worlds and thus undergird the formulation of truly novel hypotheses. As brilliant as Galileo was, without advances in optics, he would not have observed the four moons of Jupiter that undergirded new models of the solar system.

Crucial to our better understanding moving forward is the clue that many psychiatric disorders run strongly in families. Based on family and twin studies, autism, schizophrenia, bipolar disorder, and attention-deficit/hyperactivity disorder are among the most heritable of all common medical disorders. Of course, genes are not fate: Both chance and specific environmental factors play significant roles as well. The substantial contribution of genes to risk of these disorders does not, however, make gene discovery simple. We now know that psychiatric disorders result from the interaction of hundreds of genes—with different combinations of genes contributing to risk in different families. Recently we have learned that different disorders such as schizophrenia and bipolar disorder share many risk genes but also have unshared risk genes.[7] Similar patterns are emerging across all of medicine.

Given the complexity and our ability to identify the many risk-associated genetic variants—some common in populations, others very rare—it has proven necessary to study tens of thousands of patients and to compare them with roughly equal numbers of healthy individuals. Such an approach was simply not feasible until modern genomic technologies made it possible to determine genotypes (to use tools such as gene chips to identify specific variants in particular places in the genome) in an inexpensive and highly comprehensive manner and to sequence DNA rapidly, accurately, and cheaply. Although steady improvements in the technologies continue, the cost of DNA sequencing has declined approximately 1 millionfold over the last decade. Just over a decade ago, at the time of the human genome project, the cost of determining each "letter" in the genetic code was about 1 dollar; given that there are 3 billion such letters (or base pairs) in each human genome, the cost was $3 billion. The cost today of sequencing most stretches of DNA can be as low as $.07 for 1 million base pairs. As a result, we have gone from knowing about a handful of likely risk-associated genetic loci for schizophrenia in 2008 to approximately 75 by the end of 2012. As larger population samples are collected, progress is accelerating in the genetic dissection of schizophrenia, bipolar disorder, and autism.

A list of risk-associated genes does not guarantee an understanding of disease or of new therapies. One exciting recent development is the emerging recognition that genes involved in schizophrenia, bipolar disorder, and autism do not represent a random sample of the genome. Rather, the genes are beginning to coalesce into identifiable biochemical pathways and components of familiar neural structures. More excitement comes from the finding that a large number of risk-associated genes in autism, schizophrenia, and bipolar disorder code for proteins involved in the structure and function of synapses.

Our best hope is that the genetics will unfold over the next several years, due to the efforts of large international consortia that have formed to recruit and to study patients. As genetic clues accumulate, scientists are devising new ways to investigate their neurobiological functions and dysfunctions. One interesting development is to use stem cell technologies to complement the use of laboratory animals with human neurons engineered from skin cells of healthy subjects and from patients. The leading approach is to take a small skin biopsy from the arms of volunteers and to transform skin fibroblasts into neural progenitors and into neurons. Genetic engineering can then be used to add risk-causing mutations to "healthy" neurons and to reverse risk mutations in patients' neurons. But it is still early in this new field, and it is not yet possible to engineer the specific kinds of neurons implicated in schizophrenia by postmortem studies.

This barrier is likely to fall soon. Whether or not engineered neurons or human neural circuits on a chip prove to be good systems for studying gene function, researchers will make substantial efforts to turn genetic clues into ideas for therapeutics. Many researchers hope that such efforts will help attract the pharmaceutical industry back to psychiatry by demonstrating new paths to treatment development. The emerging genetic results may be the best clues we have ever had to the etiology of psychiatric disorders. If other areas of medicine can guide us, there is enormous promise in deprioritizing existing drugs and old-fashioned animal-based assays as investigative tools and instead focusing on actual disease mechanisms identified by genetics. Technology has only recently begun to make this possible.

5

Sound the Alarm
Fraud in Neuroscience

By Stephen G. Lisberger, Ph.D.

Steve Lisberger, Ph.D., is the George Barth Geller Professor of Neurobiology and the chair of the Department of Neurobiology at Duke University, and an investigator of the Howard Hughes Medical Institute. At University of California, San Francisco from 1981 to 2012, Lisberger was a professor of physiology, the founding director of the W.M. Keck Foundation Center for Integrative Neuroscience, and a co-director of the Sloan-Swartz Center for Theoretical Neurobiology. Lisberger won the Young Investigator Prize from the Society for Neuroscience in 1986 and the Bernice Grafstein Prize for achievements in mentoring women in neuroscience in 2011. A former section editor and a senior editor for the *Journal of Neuroscience*, he is chief editor of *Neuroscience*, the flagship journal of the International Brain Research Organization. Lisberger is the treasurer-elect of the Society for Neuroscience. His laboratory studies how we learn simple motor skills, and how we use what we see to guide our movements.

We expect scientists to follow a code of honor and conduct and to report their research honestly and accurately, but so-called scientific misconduct, which includes plagiarism, faked data, and altered images, has led to a tenfold increase in the number of retractions over the past decade. Among the reasons for this troubling upsurge is increased competition for journal placement, grant money, and prestigious appointments. The solutions are not easy, but reform and greater vigilance is needed.

DO SCIENTISTS CHEAT? If so, what are their motives? Are there just a few occasional offenders, or does the recent spate of scientific misconduct cases represent the tip of an iceberg? How much of a problem is subconscious cheating? How can we police ourselves so that fraud decreases, even as the pressures on scientists grow?

These are questions I ask myself as the chief editor of an international journal, the chair of a neurobiology department, and the principle investigator of a systems neuroscience laboratory.

I've been fortunate to sidestep any problems with fraud in my own department and laboratory, but in the world outside my own, I have witnessed several seemingly undeniable cases of fraud. I also have seen a larger number of cases that seemed like fraud but could have been errors, plenty of cases that aroused suspicion, and many examples of simple errors that were not fraud but resembled it. Finally, I have received accusations of misconduct that turned out, upon investigation, to be misunderstandings or pure inventions. Even though cheating seems to be a growing problem in science, we need to be cautious about proclaiming someone guilty of fraud before all the evidence is available.

In my opinion, it is likely that the field of neuroscience is detecting only the tip of the fraud iceberg. Even though most scientists conduct their research impeccably, there is more misconduct than journal editors and the scientific community detect. This is mainly because cheating can be difficult to uncover.

Recent news stories in scientific and lay press highlight an elevated number of high-profile findings that have proved difficult for others to replicate, numerous retractions of published papers, and a few widely known examples of blatant misconduct. Of course, failure to replicate another scientist's findings can result from sloppiness or divergent techniques, and retraction can occur because of an honest error. I want to believe that all the results from my field—indeed, from my own laboratory—are genuine, and that suspicious situations come from honest errors. I trust my colleagues, students, postdocs, and research staff implicitly, and I believe that they are genuinely interested in the truth. But how can I or anyone else know for sure?

The Nature of Fraud

Some kinds of fraud cause great concern. Some scientists create data that support their hypotheses, and others adjust data so that the results are statistically significant or are just cleaner and more compelling.

Some scientists appropriate others' ideas or borrow promising approaches after seeing them in a communication at a meeting, or in a grant or paper they have been asked to review. In science, ideas tend to evolve in parallel, at the same times, in the minds of different groups within a subfield, and it can be difficult to assign ownership. But I know of situations where a senior scientist appears to have stolen the ideas of a young scientist and used the ideas to his or her own advantage. I am sure it happens more often than meets the eye, even though successful senior scientists should have enough ideas of their own and do not need to steal them.

I also often see misconduct that we as a field must label as wrong but that probably results from misunderstanding, frustration, and communication failures. We can hope to reduce disagreements about authorship after a paper has been submitted or published via training in the responsible conduct of science and the trend of identifying the contribution of each author in a footnote. Proper training also should alert scientists to the prohibitions against reproducing published material without prior permission and against submitting a paper to two journals at the same time. Scientists can minimize these kinds of misconduct by better educating our young scientists about the rules of acceptable ethical conduct.

True, plagiarism is fraud. But most plagiarism occurs because an author comes from a culture where copying is a sincere form of flattery, or because an author simply doesn't understand the importance of rephrasing an idea in one's own words. As an editor, I take a generous view toward minor examples of plagiarism that I detect during review, and I advise authors to rewrite in their own words. Also, I do not regard as plagiarism the reuse of one's own words from a previous paper to describe routine methods. Still, I know that others do, and I prefer to see the material rewritten in brief form with a reference to the more authoritative description elsewhere.

Detection of Fraud

One reason for the proliferation of fraud is that it's easy to hide and difficult to detect. In my role as an editor, instances of potential misconduct generally come to me quite late in the review process. Accusations usually occur in the form of an e-mailed complaint from someone who is closer to the situation and/or more expert than I in the particular field. I estimate that I receive complaints about scientific misconduct for about 2 percent of submissions. I believe that the rate is even higher in some other journals.

Fraud can go unnoticed through much of the research process because so much of science is solitary, and we are frequently alone with our data. Laboratories can operate quite independent of the world around them. For many institutions, critical procedures and analyses are performed in shared research cores that are run by individual departments or by the institution for the common good, rather than by an individual laboratory. There may be reduced, even inadequate, oversight of these shared research cores. It would be easy to miss sleight of hand from someone who seems to have so-called golden hands. We should always remember the adage, "If it seems too good to be true, then it probably is too good to be true."

For reasons that I do not quite understand, data that are simply made up can escape the review process and not come to light until a paper appears on a journal's website and is read by an expert. Why do our reviewers fail to recognize a single image that reappears in multiple figures to show different results, or error bars that are all the same length across an entire graph? Could we create a rigorous process that would detect this kind of occurrence, determine whether it results from an honest error or from malicious misconduct, and correct it?

Fastidiously kept notebooks and experimental logs are measures that should enable data provenance and help to prevent data fraud, but cheaters will always find ways to cheat in spite of structure designed to prevent cheating. I realize it would be a big change of culture, but data fraud would be reduced—and the quality of the entire scientific literature improved—if we established requirements for publishing not just your paper, but also your data. Modern technology would allow authors to catalog and store data in a way that tracks all changes to the record, and to link the data to the figure derived from them. We only have to decide that this is the direction we need to go in.

I also wonder why authorship complaints frequently arise very late in the publication process. How can a principal investigator shepherd a paper all the way to publication without anyone along the way noticing that the paper is based on the work of a former Ph.D. student who is not listed as a co-author? In contrast, duplicate submissions often come to light very quickly. When authors submit a paper to two journals at the same time, scientists solicited to review the paper are very quick to point out that they were asked to review a seemingly identical paper for another journal.

It is probably true that the more serious the fraud, the harder it is to detect. As a result, we should be careful not to punish the minor offenses too harshly, or pursue them too vigorously, when the really serious ones may be going unpunished.

My wish is that cases of potential fraud or scientific misconduct in my department or laboratory come to my attention quite early, presumably in the form of a report from an innocent bystander who notices something that doesn't seem right. To aid this process, and to reduce fears of retaliation, I have placed an anonymous whistle-blower form on my department's website. It is part of my job as department chair and principal investigator to understand these situations, to confront them, and to resolve them. I imagine that other chairs see their jobs the same way.

Incentives for Fraud

Scientists cheat because fraud can be rewarding. Generally speaking, scientists understand that the goal of research is to learn about the truths of the world, and most scientists would not cheat simply to achieve greater rewards. But the evidence indicates that some scientists do.

I think that fraud has increased since I came into scientific research 40 years ago, as the challenges of running a successful research laboratory, obtaining funding, and publishing papers likewise have increased. In the not-so-recent past, we did not have cutthroat competition to publish in the most prestigious journals as we do today, and grant funding flowed freely. There was enough reward to go around. The life of a scientist was relatively simple, so there were fewer incentives to cheat. While we cannot "rerun the tape" (credit to the late Stephen J. Gould), I suspect that my own career path would have been a hundred times more competitive and stressful now than it was back then.

So what can we do to return to how things once were? The rewards of science rise out of publications, but simply publishing does not guarantee success. We are increasingly judged according to *where* we publish rather than *what* we publish. Remarkably, we are ranked in proportion to the number of citations garnered by the *other* papers in the journals that contain our papers (the impact factor). Some organizations decide promotions and grant applications on the basis of the impact factors of the journals that publish a scientist's papers; so rewards come from being published in the journals with the highest impact factors. As a result, the perception of the need for this kind of reward runs strong and deep.

A journal acquires a high impact factor by following a very selective review process. Because it publishes only a small fraction of the papers it receives, publication becomes highly competitive. Yes, the papers in that journal truly are better, on average, than are papers in other journals. But the field elevates the perception of the papers in the journal to such an extent that publication in that journal becomes a very high reward. Some people will stop at nothing—including cheating—to produce a paper that is exciting enough and seemingly reliable enough for such an elevated publication status.

The situation becomes more dangerous when young people start to believe—as they often do—that they can expect to get a postdoc position or a job only after they publish at least a couple of papers in the highest-ranking journals. Sadly, this perception has a kernel of truth to it. If an institution advertises for a faculty job and receives 350 applications, then it is only natural to screen for people whose work has been published in the very top journals. That kind of search might overlook someone of depth, creativity, and substance who values discoveries themselves more than the world's perception of the discoveries.

I do not think that the high-profile journals or their editors are at fault. They are doing their jobs.

We, the scientists, are at fault. We need to change how we evaluate our colleagues. We need to read their papers from cover to cover and teach our students to do so as well. We need to reject the idea that riffling through a table of contents and reading abstracts constitutes "keeping up with the literature." We need to judge scientists and their papers by how much they have truly pushed their field forward rather than by where the papers appear. We need to celebrate the excellent science that appears in specialist journals as much as we do the papers in high-profile publications. In recruiting, we need to look at the substance of an applicant's publications rather than measuring the reputations of the journals that published the papers.

And what about grant money? Doesn't the pursuit of research funding provide potential rewards for cheating? It absolutely does. A recent high-profile misconduct case allegedly involved falsification of preliminary data for a grant application. I doubt this is the first (or the last) time such misconduct has occurred. It would be glib and simple to say that the incentive of potential grant money could be reduced (but not removed) if pay lines were much better at the National Institutes of Health.

An Insidious Form of Fraud

As the chief editor of a journal, I find it especially challenging to evaluate one type of subtle cheating. When a reviewer is evaluating a paper for publication, there is an opportunity for conscious or, worse, subconscious bias in the review. If the paper has been submitted to a high-profile journal, the reviewer might be thinking, "This paper could take my paper's place in this journal. I need to find reasons to make sure that doesn't happen." It is possible to delay a competitor simply by taking a long time to complete and submit a review. It is easy for reviewers to delay publication or to push a paper down in the journal food chain by asking for more experiments. The advent of "supplementary material" has made it easier for reviewers to ask for more data. Supplementary material is not part of the actual paper, but consists of additional figures, graphics, and tables that are maintained on a journal's website and are available only on the internet. I advocate limiting supplementary material to formats such as audio and video clips that cannot easily be embedded in a PDF, thus obviating this particular device that is now available to reviewers. Supplementary figures and text have been disallowed at least in *Neuroscience* and the *Journal of Neuroscience,* and limited in some other journals. People have complained, of course, but the quality of the papers published by these journals has not suffered.

As an editor, I always wonder whether a reviewer's request for more experiments is asking for a fundamental piece of the story, or requiring the authors to go way beyond any reasonable scope for the paper they have presented. As the chair of a study section in the past, I frequently had to remind my reviewers that they were to evaluate the grant they had been given, not try to rewrite it. As editors and reviewers, I think we need to try to evaluate the paper we've been given, not try to transform it. I do not know whether this problem will be resolved by new journals that promise an efficient review process and do not make requests for additional experiments that expand the scope of the paper, but it is an experiment worth trying.

In regard to subconscious bias, I suggest that there may be a ton of subconscious cheating going on in how experiments are done, how data are selected and analyzed, what is and is not told to the Principal Investigator of the lab, and how the message of a paper is massaged. My colleagues who are psychiatrists assure me that the subconscious exists and that it plays a key role in our actions. Subconscious misconduct is one of the most serious issues we face, and I suspect there is much more of it than there is of conscious, intentional cheating.

What Can We Do?

There are parallels between the situation with scientific fraud today and the situation with animal use (and abuse) 30 years ago: clear examples of inappropriate treatment of animals then and of scientific fraud now; an unknown number of undetected abuses then and the likelihood of more than a few undetected cases of fraud now; many instances of substandard treatment of animals then and of possibly inappropriate scientific conduct now; and an overwhelming majority of scientists who were using animal subjects appropriately then and who are conducting science impeccably now. Unfortunately, data fraud may be more difficult to spot than is animal abuse.

To reduce scientific fraud, perhaps we can learn some lessons from the scientific community's responses to the actions of animal activist organizations in the 1980s and 1990s. Those organizations, through actions of questionable appropriateness and (sometimes) legality, mobilized the scientific community to police itself. As a result, we now have detailed protocols that outline procedures to ensure the welfare of our animal subjects. We follow the protocols, and there is a structure for regulatory oversight that has largely (but not completely) eliminated abuses. Adherence to regulations, laws, and protocol is a key part of the ethics training of young scientists. We have anonymous whistle-blower avenues for anyone who has a concern. And the issue is discussed openly at venues as large as the annual conferences of our major professional societies and as small as individual lab meetings. While the process has been unpleasant at times, our treatment of animal subjects has improved dramatically.

In the realm of scientific fraud, there are now activist groups such as Reaction Watch and Clare Francis, as well as software that helps detect plagiarism. They have raised awareness of the issue for some of us. Thus, the first step is to broaden awareness of the issues of fraud so that the discussion occurs in all institutions and all laboratories, not just in those that have been touched by an incident of serious scientific misconduct.

Another strategy is to publicize cases of fraud ourselves rather than leaving it to outside activists. Perhaps we need a stronger and more visible regulatory structure to detect fraud earlier in the steps required to complete and publish a research project.

Journals and institutions alike are stakeholders. Both stand to lose if fraud continues, and both should play proactive roles in detecting and thereby removing the incentives for fraud. We, as scientists, should reduce our admiration of "high-profile" publications and evaluate scientists for jobs, promotions, grants, and tenure on the basis of what they have done rather than where their research has been published.

Finally, we should talk about misconduct more often and more deeply. Subconscious and conscious misconduct needs to be discussed in lab meetings, faculty meetings, ethics courses, and national meetings. By putting fraud under the light and developing a strong structure for its detection, we can reduce it dramatically, even if we will never be able to eliminate it altogether. And we need to remember that although fraud may be more prevalent than we think, most scientists conduct their research irreproachably. As always, we need to be careful not to assume fraud has occurred just because there's been an accusation. Investigation often reveals that an error, a misunderstanding, or nothing at all has occurred.

6

Inside the Letterbox

How Literacy Transforms the Human Brain

By Stanislas Dehaene, Ph.D.

Stanislas Dehaene, Ph.D., directs the INSERM-CEA Cognitive Neuroimaging unit, located near Paris, and is the author of *The Number Sense* (1999) and *Reading in the Brain* (2009), both broadly acclaimed and translated. In 2005, at the age of 40, he was elected to the French Academy of Sciences and as a professor of the *Collège de France* in Paris, which created a new chair of Experimental Cognitive Psychology for him. Dehaene has received several international prizes, including the James S. McDonnell Centennial Fellowship, the Louis D. Foundation Prize, and the Heineken Prize for Cognitive Science. A March 2008 profile in *The New Yorker* ("Numbers Guy") called him "one of the world's foremost researchers," "completely pioneering," and "a scanning virtuoso" in the field of numerical cognition.

Few issues are as important to the future of humanity as acquiring literacy. Brain-scanning technology and cognitive tests on a variety of subjects by one of the world's foremost cognitive neuroscientists has led to a better understanding of how a region of the brain responds to visual stimuli. The results could profoundly affect learning and help individuals with reading disabilities.

⟨━━━━⟩

ALTHOUGH I FIND THE DIVERSITY of the world's writing systems bewildering, there is a striking regularity that remains hidden. Whenever we read—whether our language is Japanese, Hebrew, English, or Italian—each of us relies on very similar brain networks.[1] In particular, a small region of the visual cortex becomes active with remarkable reproducibility in the

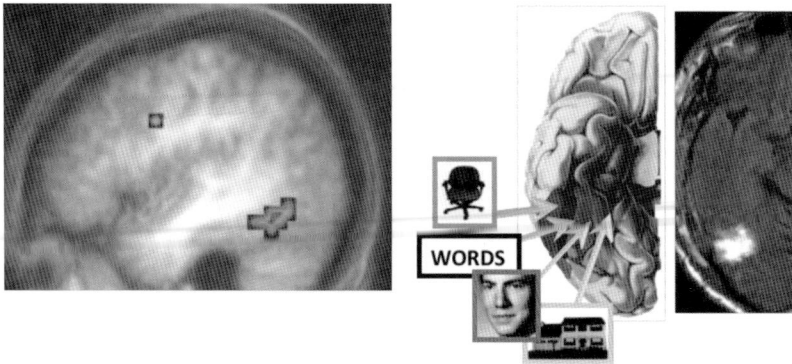

Figure 1. *The visual word form area—the brain's letterbox—is a small region of the human visual system that systematically activates whenever we read. It shows a stronger activation to words than to many other categories of visual stimuli, such as pictures of objects, faces, or places. In all of us, it is systematically located at the same place within a "mosaic" of ventral preferences for various categories of objects. And, if it is destroyed or disconnected, as in the patient whose brain scan is shown at right, we may selectively lose the capacity to read.*

Left image is modified from Dehaene, S., Naccache, L., Cohen, L., Le Bihan, D., Mangin, J. F., Poline, J. B., & Rivière, D. (2001). Cerebral mechanisms of word masking and unconscious repetition priming. *Nat Neurosci,* 4(7), 752-8. Middle image modified from Ishai, A., Ungerleider, L. G., Martin, A., & Haxby, J. V. (2000). The representation of objects in the human occipital and temporal cortex. *J Cogn Neurosci,* 12 Suppl 2, 35-51. Right image from Gaillard, R., Naccache, L., Pinel, P., Clemenceau, S., Volle, E., Hasboun, D. Cohen, L. (2006). Direct intracranial, FMRI, and lesion evidence for the causal role of left inferotemporal cortex in reading. *Neuron,* 50(2), 191-204.

brains of all readers (see figure 1). A brief localizer scan, during which images of brain activity are collected as a person responds to written words, faces, objects, and other visual stimuli, serves to identify this region. Written words never fail to activate a small region at the base of the left hemisphere, always at the same place, give or take a few millimeters.[2]

Experts call this region the visual word form area, but in a recent book for the general public,[3] I dubbed it the brain's "letterbox," because it concentrates much of our visual knowledge of letters and their configurations. Indeed, this site is amazingly specialized. The letterbox responds to written words more than it does to most other categories of visual stimuli, including pictures of faces, objects, houses, and even Arabic numerals.[4]

Its efficiency is so great that it even responds to words that we fail to recognize consciously—words made subliminal by flashing them for a fraction of a second. Yet it performs highly sophisticated operations that are indispensable to fluent reading. For instance, the letterbox is the first visual area that recognizes that "READ" and "read" depict the same word by representing strings of letters invariantly for changes in case, which is no small feat if you consider that uppercase and lowercase letters such as "A" and "a" bear very little similarity. Furthermore, if it is impaired or disconnected via brain surgery or a cerebral infarct (type of stroke), the patient may develop a syndrome called pure alexia. He or she will be unable to recognize even a single word, as well as faces, objects, digits, and Arabic numerals. Yet many of these patients can still speak and understand spoken language fluently, and they may even still write; only their visual capacity to process letter strings seems dramatically affected.

The brain of any educated adult contains a circuit specialized for reading. But how is this possible, given that reading is an extremely recent and highly variable cultural activity? The alphabet is only about 4,000 years old, and until recently, only a very small fraction of humanity could read. Thus, there was no time for Darwinian evolution to shape our genome and adapt our brain networks to the particularities of reading. How is it, then, that we all possess a specialized letterbox area?

Reading as Neuronal Recycling

Resolving this paradox requires thinking about the state of the brain prior to literacy. According to a theoretical proposal called the neuronal recycling hypothesis, which I introduced with colleague Laurent Cohen a few years ago, the human brain contains highly organized cortical maps that constrain subsequent learning.[5] We should stop thinking of human culture as a distinctly social layer, free to vary without bounds, independent of our biological endowment. On the contrary, new cultural inventions such as writing are only possible inasmuch as they fit within our preexisting brain architecture. Each cultural object must find its neuronal niche—a set of circuits that are sufficiently close to the required function and sufficiently plastic to be partially "recycled." The theory stipulates that cultural inventions always involve the recycling of older cerebral structures that originally were selected by evolution to address very different problems but manage, more or less successfully, to shift toward a novel cultural use.

How can this view explain why all readers possess a specialized and reproducibly located area for a recent cultural invention? The idea is that the act of reading is tightly constrained by the preexisting brain architectures for language and vision. The human brain is subject to strong anatomical and connectional constraints inherited from its evolution, and the crossing of these multiple constraints implies that reading acquisition is channeled to an essentially unique circuit.

As far as *spoken* language is concerned, all humans rely on a highly determined network of left superior temporal and inferior frontal regions. This system is so reproducible that it is found in any individual. Indeed, it can already be evidenced with functional magnetic resonance imaging (fMRI) in 2-month-old babies when they listen to short sentences in their mother tongue[6]—and even at this early age, the language network is already lateralized to the left hemisphere in most subjects. Somewhat unsurprisingly, the lateralization of this spoken language system constrains the lateralization of reading. The visual word form area is systematically lateralized to the same hemisphere as spoken language: It typically falls in the left hemisphere in most people, but it shifts to the right ventral occipitotemporal region in those rare subjects with right-hemisphere language.[7] Presumably, this is because the ventral temporal visual cortex lies very close to the temporal language areas that encode spoken words and speech sounds, thus allowing this region to serve as a visual interface for reading. This optimally short circuit may allow a good reader to convert back and forth between letters and sounds with minimal transmission time.

Many other constraints probably conspire to mark the precise location of the brain's letterbox area. For instance, this region always falls within the part of the visual cortex that receives inputs from the high-resolution region of the retina called the fovea.[8] Thanks to this projection pattern, this cortical sector can discriminate very small shapes—obviously a highly desirable feature when considering that reading involves resolving the often minuscule differences that distinguish similar letters, especially in small print.

Another important constraint is that neurons in the ventral visual pathway often respond to simple shapes, such as those formed by the intersections of contours of objects.[9] Even in the macaque monkey, the inferotemporal visual cortex already contains neurons sensitive to letter-like combinations of lines such as T, L, X, and ★.[10] The ventral visual system seems to favor those shapes because they signal salient properties of objects that tend to be robustly invariant to a change in viewpoint. For instance, a T frequently signals occlusion of one object by another: The vertical contour disappears behind the horizontal one, indicating that the latter marks the contour of an opaque surface that lies in front of the former. Across evolution and development, the visual system of all primatesmacquires a whole "alphabet" of such shape primitives, presumably because they allow us to immediately encode any new shape that we encounter.

The visual word form area seems to arise from this preexisting neuronal alphabet: A subset of the shape-recognition system specializes in the shapes of letters. In my book *Reading in the Brain,*[11] I further speculate that in all of the world's cultures, scribes in generation after generation progressively selected their letters and written characters to closely match the set of shapes that were already present in the brains of all primates and, as a result, were easy to learn. This hypothesis is corroborated by a large-scale analysis of the world's writing systems.[12]

Writing systems do vary in their "grain size": the linguistic units that are marked in writing vary from phonemes (in our alphabet) to syllables (in Japanese Kana notation) or even entire words or morphemes (as in Chinese).

However, visually speaking, they systematically make use of the same set of shapes, precisely those that abound in natural visual scenes and tend to be internalized in the ventral visual cortex.[13] All writing systems seem to rely on the set of shapes to which our primate brain is already highly attuned—living proof that culture is constrained by brain biology.

In the end, the neuronal recycling hypothesis leads me to believe that we are able to learn to read because within our preexisting circuits, there is one that links the left ventral visual pathway to the left-hemispheric language areas. This circuit is already capable of recognizing many letter-like shapes, and it possesses enough plasticity, or adaptability, to reorient toward whichever shapes are used in our alphabet. But this region is highly constrained, explaining why we all use the same cortical circuit and the same set of basic shapes when we read.

Scanning the Illiterate Brain

In the past eight years, my research has put the predictions of the neuronal recycling hypothesis to the test. One of the predictions, which I find particularly interesting, is the notion of a cortical competition process. The model predicts that, as cortical territories dedicated to evolutionarily older functions are invaded by novel cultural objects, their prior organization should slightly shift away from the original function (though the original function is never entirely erased). As a result, reading acquisition should displace and dislodge whichever evolutionary older function is implemented at the site of the visual word form area.

To test this idea, we needed to figure out the organization of the human brain prior to reading: the architecture of the illiterate brain. Indeed, at present, our knowledge of the human brain is highly biased. Most of it comes from neuroimaging experiments in highly educated students. My laboratory recently took a different approach by launching a major investigation into the brains of uneducated illiterate adults.[14] This collaborative effort, which involved colleagues in Belgium, Portugal, and Brazil, aimed to obtain a detailed map of the visual, auditory, and language areas in people who never had the chance to learn to read. By comparing these areas with those of literate adults, we hoped to illuminate how literacy transforms the brain.

We recruited adults around the age of 50. They held jobs and were integrated into society but had not attended school during their youth. Ten remained utterly illiterate, unable even to recognize most letters. Twenty-one were "ex-illiterates," meaning that although they hadn't attended school, they had later received adult alphabetization classes and, consequently, had achieved a modest and variable level of literacy. We compared them to 32 literate adults, some of whom came from the very same socioeconomic groups as the illiterates did.

The results demonstrated a large impact of reading acquisition on the brain (see figure 2). In the visual system, the visual word form area was massively affected. At its usual site in the left lateral occipitotemporal sulcus, the activation evoked by a string of letters was directly proportional to the subjects' reading score, to such an extent that one could predict a large proportion of the variance in the number of words they could read per minute, simply by measuring their brain activity.

We also observed unsuspected changes within the brain areas that implement the initial stages of visual recognition. With literacy, lateral occipital areas increased their activation not just to words, but also to all sorts of stimuli (faces, houses, checkerboards), suggesting that learning to read had refined the capacity to recognize any

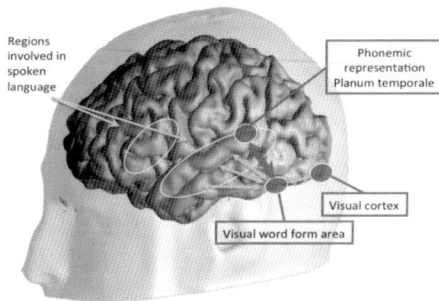

Figure 2. A vast brain circuit is transformed when we learn to read. As described in the main text, all of the regions shown in the regions with dots increase their activation and specialization during the acquisition of literacy. Furthermore, a massive bundle of connections linking the ventral visual areas of the left hemisphere with the superior temporal regions involved in phonological coding is also reorganized. As a result of those changes, we gain the ability to access the spoken language system through vision.

Courtesy of Stanislas Dehaene

picture. In fact, even the primary visual area V1, which is the first point of entry of visual signals into the cortex, was enhanced. Relative to illiterates, trained readers showed enhanced fMRI responses to horizontal than to vertical checkerboards. Intensive training with horizontally presented words had obviously led to a refinement of the corresponding region of the visual field, in the part of the eye that is indispensable for reading fine print. Behaviorally, the literates performed much better than the illiterates at a contour integration task that is thought to depend heavily on the earliest cortical stage of vision, the primary visual area V1 and its horizontal connections.[15] In brief, learning to read seems to render even the lower-level processes of early vision more accurate and more efficient.

Most crucially for the neuronal recycling hypothesis, however, was our determination that reading did not merely have a positive effect on the brain. Exactly as predicted, we also observed a small but significant cortical competition effect, precisely at the site of the letterbox area. For the first time, our study revealed which shapes triggered a response at this site prior to learning to read. In illiterates, faces and objects caused intense activity in this region—and, strikingly, the response to faces diminished with literacy. It was highest in illiterates, and quickly dropped in ex-illiterates and literates. This cortical competition effect, whereby word responses increased while face responses decreased, was found only in the left hemisphere. In the symmetrical fusiform area of the right hemisphere, face responses increased with reading. Thus, at least part of the right-hemisphere specialization for faces, which has been repeatedly observed in dozens of fMRI studies, arises from the fact that these neuroimaging studies always involved educated adults. Obviously, the acquisition of reading involves the reconversion of evolutionary older cortical territory, and text competes with faces for a place in the cortex.[16]

We also replicated this finding in children.[17] When scanning 9-year-olds who were good readers versus dyslexic readers, we found two interesting differences in the ventral visual pathway (see figure 3). The dyslexic readers not only showed weaker responses to written words in the left-hemispheric visual word form area, but also showed weaker responses to faces in the right-hemispheric fusiform face area. Thus, the acquisition of reading seemed to induce an important reorganization of the ventral visual pathway,

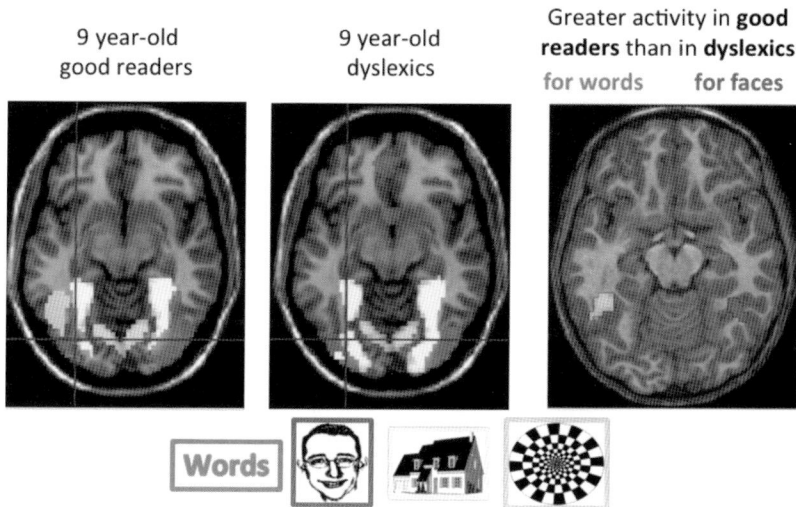

9 year-old
good readers

9 year-old
dyslexics

Greater activity in **good
readers** than in **dyslexics**

for words   for faces

Words

Figure 3. The organization of visual areas differs in good readers and in dyslexics (redrawn from Monzalvo et al., 2012). Good readers show a well-developed visual word form area (shown in dark gray), a region that responds to written words more than to other categories of visual stimuli (faces, places, or checkerboards). Dyslexics show no such specialization for written words, and also exhibit a much weaker activation to faces in the right hemisphere. The evidence suggests that literacy involves a specialization of the visual word form area in the left hemisphere and, as a result, the displacement of face responses to the right hemisphere—two processes that fail to occur in young dyslexics.

From Monzalvo, K., Fluss, J., Billard, C., Dehaene, S., & Dehaene-Lambertz, G. (2012). Cortical networks for vision and language in dyslexic and normal children of variable socio-economic status. *Neuroimage*, 61(1), 258-74. doi:10.1016/j.neuroimage.2012.02.035

which displaces the cortical responses to turn away from the left hemisphere and more toward the right. This displacement is presumably because the features that are most useful for letter recognition (configurations and intersections of lines) are incompatible with those that are useful for faces, so that one pushes the other away.

There is no reason to worry, however. In our research so far, we have been unable to uncover any negative behavioral consequences of literacy on face-recognition abilities. Although face recognition is displaced in the cortex, it seems to be just as efficient in literate people as in illiterate people. In fact, it may even be more efficient in literate people. In a test of holistic perception, where subjects were asked to compare the top halves of two faces while avoiding any interference from their bottom halves, literate people outperformed illiterate people, suggesting that the former had learned to focus their attention better and in a more flexible manner. Many of us complain about the difficulty of recognizing faces and retrieving people's proper names in everyday life, but the true culprit is unlikely to be an excess of reading.

Unlearning Mirror Invariance

Another feature of the reading system—the need to discriminate mirror-symmetrical images—places an unusual constraint on the visual system. Our alphabet comprises pairs of letters such as "p" and "q" or "b" and "d," which are similar except for a left-right inversion. In order to read fluently, we must learn to discriminate these letters perfectly, because they point to different phonemes. Remarkably, however, in its preliterate state, the inferotemporal cortex of all primates generalizes across such mirror images and treats them as two views of the same object.[18] Indeed, adults, children, and even infants immediately recognize an object regardless of whether its left or right profile is seen. In macaque monkeys, inferotemporal neurons spontaneously generalize across mirror images. When a monkey is trained with a shape in a specific orientation, the neurons that specialize for this shape generalize to its mirror image without any further training.

My colleagues and I reasoned that mirror invariance was a perfect test of the neuronal recycling hypothesis. Here was a preexisting competence, clearly present in the brains of all primates, and yet if our hypothesis was correct, it had to selectively disappear in the course of reading acquisition. Mirror invariance had the potential to become the "panda's thumb" of reading—a vestigial trait, inherited from our evolution in a natural world where most objects maintain their identity across a left-right inversion. Its presence prior to literacy and its disappearance during reading acquisition would imply that learning to read involves the recycling of a preexisting circuit that never evolved for reading but willy-nilly adapts to this novel task.

There is, in fact, much evidence to support this theory. Early on during reading acquisition, most children generalize across mirror images to such an extent that they are able to read and write their first words independent of orientation. Just ask any 5- or 6-year-old child to write his or her name next to a dot located near the right side of the page. Most of them will unhesitatingly solve the problem by writing from right to left.[19] This mirror competence slowly disappears during reading acquisition, but it remains present in illiterate people who, contrary to literates, exhibit no cost at all in recognizing a learned object in mirrored form. Indeed, illiterate people find it extraordinarily hard not to see "b" and "d" as identical shapes. The breaking of this spontaneous mirror invariance is one of the important outcomes of literacy.

Using fMRI, we found that the visual word form area is, once again, the site of this adaptation to reading.[20] When we used an fMRI priming technique in adult readers, we observed no mirror invariance to words in the brain's letterbox area. This region easily recognized that "radio" and "RADIO" were one and the same word, but it did not label a word and its mirror image as identical ("radio" and "oidar"). It did so, however, whenever pictures were presented (even when they were visually just as simple as a single letter). Surprisingly, our fMRI study revealed that the visual word form area is the place of the visual system with the strongest mirror invariance for pictures. No wonder, then, that young readers experience special difficulties with mirror letters such as "b" and "d": They are learning to read via the precise site of the cortex that treats these letters as identical. This difficulty is not connected to dyslexia; it is due to a universal feature of the primate brain that all children possess and must unlearn. Only if mirror errors persist beyond the ages of 9 or 10 should parents and educators worry, because the prolongation of mirror errors beyond that age suggests that the unlearning processing is not proceeding normally.[21]

The unlearning of mirror symmetry should not be solely construed as a loss. As we gain literacy, we become slightly worse in judging that two mirror images are the same, but we also gain an enormous advantage, which is the capacity to distinguish them. Behavioral tests show that illiterate adults, even if their intelligence is high, find it extremely hard to distinguish nonsense shapes such as < and >, since they cannot help but see the shapes as identical.[22] Using recordings of the brain's event-related potentials in our literate and illiterate participants, we recently found that the capacity to discriminate mirror-symmetrical pseudo-words such as "iduo" and "oubi" increases with literacy. Again, literacy enhances our species' behavioral repertoire by allowing our visual system to flexibly treat shapes as identical or different, depending on context.

Literacy Enhances Speech Processing

Up to now, we focused primarily on how learning to read changes the *visual* brain. However, our study of literate and illiterate brains also uncovered a massive and positive effect of literacy on the network for *spoken* language processing. First, and perhaps trivially, in literate people but not in illiterate people, the language network of left temporal and inferior frontal regions activates very strongly and identically to written and spoken language. Thus, the acquisition of reading gives us access, from vision, to a broad and universal language processing system—the very same language circuit that is already active in 2-month-old babies. Writing really acts as a substitute for speech, and ends up activating the same areas of the brain.

Second, and most important, this network for spoken language also changes under the influence of reading. In most areas, we found decreased activation. In order to understand the very same spoken sentence, the best readers required less activity than the illiterate people. Indeed several brain areas associated with mental effort, such as the anterior cingulate, decreased in activity dramatically with literacy, confirming that reading facilitates the comprehension of complex sentences.

In one auditory area, however—the left planum temporale, located just behind the primary auditory area—we saw a strongly increased response to spoken sentences, words, and even meaningless pseudo-words. We think that this region may be the site of one of the most important competences that we acquire when we learn to read: the capacity to convert letters into sounds, or graphemes into phonemes. During infancy, the planum temporale region plays an essential role in the acquisition of the phonemes and the phonological rules that are particular to a person's mother tongue.[23] The additional changes that this region undergoes in literate people may reflect how literacy changes the phonological code. Many behavioral studies have demonstrated that alphabetization leads to a considerable improvement in phonological awareness. Readers of an alphabetic language gain explicit access to a representation of phonemes, the fundamental units of spoken language.[24] Illiterate people cannot hear the identical "b" phoneme in "back" and "cab," and they cannot selectively delete it to produce "hack" or "ca." It seems very likely that the near doubling of brain activity that we saw in the left planum temporale relates to this massive reorganization of spoken language. Literate and illiterate brains differ in the very manner in which they encode speech sounds.

A Bidirectional Sight-to-Sound Pathway

Our fMRI study of literacy revealed yet one more change in the way good readers process speech. When listening to spoken language, they alone could activate the orthographic code in a top-down manner. Indeed, we saw a massive activation at the exact site of the visual word form area whenever the good readers were asked to decide whether a spoken item (e.g., "ploot") was or was not a word in English. During the same lexical decision task, illiterate people showed no trace of any such recruitment of their letterbox system. We therefore concluded that with literacy comes the capacity to recode speech with vision; it obviously helps to consider a sound's possible spellings before deciding whether it is a word or not.

Prior behavioral results showed a huge influence of spelling on speech processing in this domain as well. For instance, English-literate people believe that there are more phonemes in "pitch" than in "rich" (just because there are more letters), and they think that by removing the sound /n/ in the spoken word "bind," one gets "bid" (the correct answer is "bide").[25] Obviously, their spoken-language judgments are deeply biased by spelling. But once again, the net effect is primarily positive, as it increases the number of redundant codes available to process speech. This expanded mental world may explain why literate people usually exhibit a much larger verbal memory than illiterate people do.

If our hypothesis is correct, then reading acquisition enhances a fast and bidirectional connection between letters and sounds—and I do mean a physical connection, a bundle of axons linking the visual word form area to the planum temporal. We reasoned that we might be able to see this putative connectivity change directly in our anatomical images of the human brain by using a technique called diffusion tensor imaging. This technique measures the microscopic organization of the "white matter" of the brain, the massive bundles of fibers that link brain areas into functional networks. One parameter in particular, called fractional anisotropy, provides a sensitive index of whether the fibers are well aligned and covered with an insulating sheet of myelin that speeds up neural transmission. To evaluate this parameter, we measured fractional anisotropy in our literate and illiterate subjects, in five different long-distance pathways of the left hemisphere.

We found that the pathway that correlated quite well with the subject's literacy was the vertical and posterior segment of the arcuate fasciculus, which links the posterior temporal lobe (including the visual word form area) with the inferior parietal lobule and posterior superior temporal regions involved in grapheme-to-phoneme conversion.[26] We also found that the anisotropy of this fiber bundle correlated with the amount of fMRI activation to written words in the visual word form area and to spoken words in the left planum temporale. This finding strengthens the hypothesis that this bundle links visual and phonological areas and participates in the grapheme-to-phoneme conversion route, whose acquisition lies at the heart of literacy.

The Next Step

Learning to read is a major event in a child's life. Cognitive neuroscience shows why: Compared to the brain of an illiterate person, the literate brain is massively changed, mostly for the better—through the enhancement of the brain's visual and phonological areas and their interconnections—but also slightly for the worse, as the displacement of the brain's face-recognition circuits reduces the capacity for mirror invariance. Once children learn to read, their brains are literally different.

Now that we understand exactly which circuits are changed by reading education, we may start thinking about how to optimize this process, particularly for children who struggle in school. Indeed, the present results dovetail nicely with prior education research, which leaves little doubt that the systematic teaching of how letters map on to speech sounds may favor the rapid establishment of the brain's visual and phonological circuits.[27] Training preschoolers with just a few hours of GraphoGame—fun software that links graphemes and phonemes—is enough to enhance the representation of letters in the cortex. In one fMRI study, after less than four hours of total training that was spread over a few weeks, the visual word form area quickly began to respond to written words in 6-year-olds.[28] By monitoring children's progress by their behavior as well as by brain imaging, we now have all the necessary tools to better understand what schools do and facilitate enhanced learning strategies.

7

Gut Feelings
Bacteria and the Brain

By Jane A. Foster, Ph.D.

Jane A. Foster, Ph.D., is an associate professor in the Department of Psychiatry and Behavioral Neurosciences at McMaster University and a member of the Brain-Body Institute, St. Joseph's Healthcare in Hamilton, Ontario, Canada. Following completion of her doctoral work at the University of Toronto, she was a postdoctoral fellow at Henry Ford Hospital in Detroit, and then a research fellow at the National Institute of Mental Health. Foster joined the McMaster faculty in 2003, where her research program investigates the brain-gut axis. Recent work from their group has established a link between gut microbiota and behavior. Ongoing work investigates the importance of gut microbiota and immune-brain crosstalk to normal brain development, and to the risk of psychiatric disorders, and eventually, how best to therapeutically target microbiota and the immune system in disease.

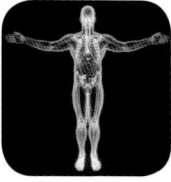

The gut-brain axis—an imaginary line between the brain and the gut—is one of the new frontiers of neuroscience. Microbiota in our gut, sometimes referred to as the "second genome" or the "second brain," may influence our mood in ways that scientists are just now beginning to understand. Unlike with inherited genes, it may be possible to reshape, or even to cultivate, this second genome. As research evolves from mice to people, further understanding of microbiota's relationship to the human brain could have significant mental-health implications.

⸻

AS A SCIENTIST, I often find myself chatting with friends and neighbors about the latest advances in neuroscience. In the past few years I have found more and more people asking about microbiota—the microorganisms that typically inhabit a bodily organ. In the last 10 years, I've been one of many neuroscientists advancing new ideas about how microbiota in the gut affects brain function. The media has taken notice as well. Recent stories on the topic include "Some of My Best Friends Are Germs" in the *New York Times Magazine* and "Gut Microbes Contribute to Mysterious Malnutrition" in *National Geographic*. In 2012, the editors of *Science* thought the research important enough to devote a special issue to the topic.

Why is the issue so fascinating? For one thing, it's heightened consciousness of how diet and nutrition impact our health. For another, it's sheer numbers. Our brains contain billions of neurons, but we less often talk about the fact that trillions of "good" bacteria are alive and well in our intestinal tracts. Remarkably, these naturally occurring, ever-present commensal bacteria in your gut may be instrumental in how your brain develops, how you behave, how you react to stress, and how you respond to treatment for depression and anxiety.

With such serious mental-health implications to consider, there is substantial buzz among neuroscientists about the bidirectional nature of these seemingly infinite relationships. I am continually impressed by the creative ways that my colleagues are making discoveries, especially in how microbiota may influence the brain and the immune system during early life. And just last week researchers at UCLA found that regularly eating yogurt with probiotics, which contain "good" bacteria, seems to affect brain functioning in women. The gut-brain axis is among the most exciting new frontiers in neuroscience.

Ups and Downs

Scientists have recognized communication between the brain and the gut for more than 100 years, with studies in the early 19th and 20th centuries showing that a person's emotional state can alter the function of the gastrointestinal (GI) tract.[1,2,3] One of the best examples is the work of William Beaumont, an army surgeon, who became known as the "Father of Gastric Physiology." In the 1830s, Beaumont, who was able to monitor gastric secretions through a fistula (a permanent opening in the stomach wall), noted an association between changing moods and gastric secretions. In the mid-to-late 1900s, research examining stress biology and its impact on human health uncovered clear connections between an individual's stress response and gut function. This classical view of top-down control—that is, the brain's ability to control gut function—is supported by evidence revealing that the brain influences body systems, including the GI tract, through neural connections of the autonomic nervous system and through humoral systems in the bloodstream. Both of these communication pathways are activated in stressful situations and influence gut function. What is exciting and new is the consideration of bottom-up control—that is, how the gut, or more precisely the microbiota in the GI tract, can influence brain function. Researchers have recently shown that the presence of gut microbiota during early development influences the brain's neural connectivity related to anxiety and depression. Gut microbiota has been linked to behavior, to stress, and to stress-related diseases. Changes in gut microbiota may influence risk of disease, and manipulating microbiota may provide novel ways to intervene in clinical situations related to mood and anxiety disorders.

The Inside Story

Normal commensal microbiota colonizes the mammalian gut and other body surfaces shortly after birth and remains there throughout an individual's lifetime. In humans, the lower intestine contains 10^{14} to 10^{15} bacteria; that is, there are 10 to 100 times more bacteria in the gut than there are somatic cells in the human body.[4] The interactions between commensal microbiota and its host are for the most part beneficial. In particular, the presence of commensal organisms is critical to immune function, nutrient processing, and other aspects of healthy physiology.[5, 6] Using the latest molecular and genetic tools, researchers have shown that several bacterial phyla are represented in the gut and that commensal populations show considerable diversity, with as many as 1,000 distinct bacterial species involved.[7] In addition, factors such as genetics, age, sex, and diet continually influence the composition and profile of an individual's microbiota.[8, 9] In healthy people, there is considerable variability in gut bacterial composition, and yet if the same person's gut bacteria are examined at different times (a few months or more apart), they hardly change.[4, 10, 12] But in stressful situations, or in response to physiological or diet changes, the microbiota profile may itself change, creating an imbalance in host-microbiota interactions. Such changes can affect the person's health.

Seeing the Light

Gut microbiota are important to healthy brain development. In particular, microbiota may influence the development of brain regions involved in our response to stress and control stress-related conditions such as anxiety and depression. In an attempt to understand these relationships, scientists manipulate gut bacteria in mice by raising germ-free mice in sterile isolators and then measuring the presence of gut bacteria. The isolator eliminates any exposure to outside air, contaminants, and commensal bacteria. Much of this work draws upon standard animal behavioral tests that measure activity, approach, and avoidance. Mice have a natural tendency to explore their environment while avoiding open and brightly lit areas. The elevated-plus maze, a behavioral apparatus that is elevated aboveground (figure 1), contains an area (with two closed and two open arms) for a mouse to explore. When a normal control mouse is placed in the maze, it tends to explore both arms but spend most of its time in the closed one. When a germ-free mouse is removed from its sterile housing conditions and placed in the maze, it tends to spend significantly more time in the open arm. This suggests that mice without gut bacteria have low levels of anxiety-like behavior, since they spend more time in the aversive area of the testing apparatus.[13, 14]

Another behavioral test uses the light-dark box, which has a dark, closed area connected to a light open area (figure 2). A normal control

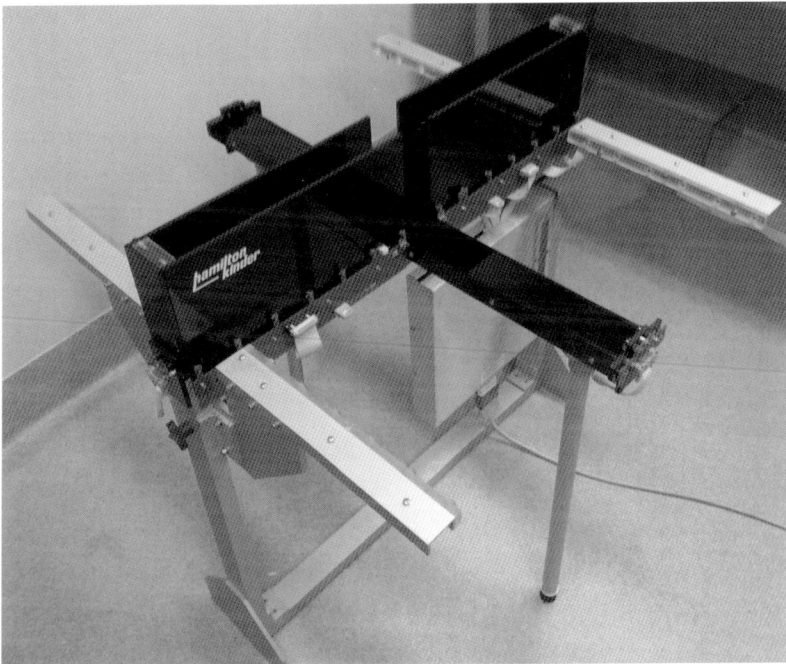

Figure 1: *The elevated-plus maze. This apparatus is elevated off the ground and consists of four black Plexiglas arms in the shape of a plus. Two of the four arms are considered closed arms as they contain black Plexiglas sides. The two open arms have small raised ledges situated along the perimeter. A test mouse is placed in the center of the apparatus and left to explore for five minutes. Infrared beams detect the placement and movement of the mouse. The elevated-plus maze was connected to a computer that collected behavioral data using MotorMonitor software.*

Photography by Sufian Odeh; image courtesy of Jane Foster

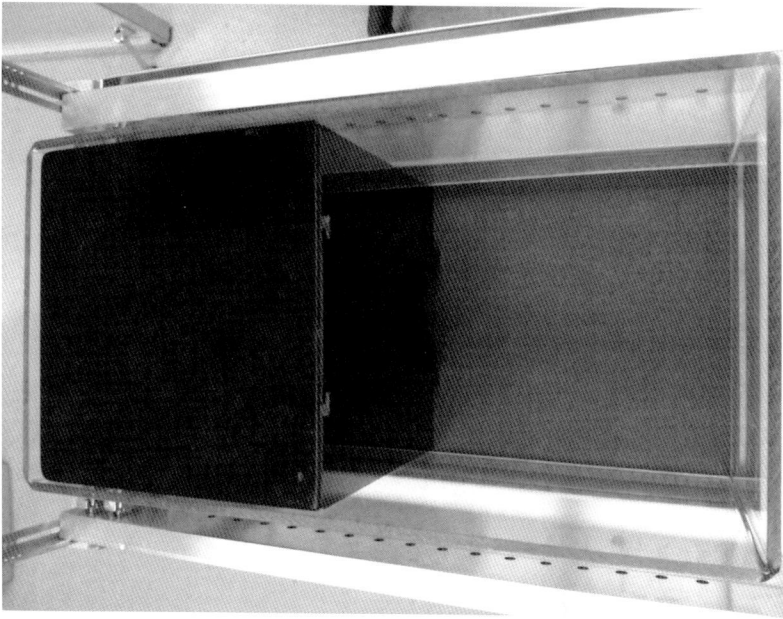

Figure 2: *Light-dark box. This apparatus—an automated infrared system containing clear a plexiglas activity chamber with a dark enclosed insert—is connected to a computer that records activity, also using MotorMonitor software. A test mouse is placed in the light side and left to explore for 10 minutes. The small opening between the chambers allows access to both the light and dark chambers for the duration of the test.*

Photography by Sufian Odeh; image courtesy of Jane Foster

mouse explores both the light and the dark chambers with a preference for the darker one. However, germ-free mice spend more time in the light side of the apparatus, again demonstrating that mice without gut bacteria have low levels of anxiety-like behavior because the light chamber is considered the aversive region in this test, and germ-free mice spend more time in the light chamber.[13, 15]

Germ-free mice have helped researchers explore whether there are particular times over a mouse's lifespan when microbiota-brain interactions are most important. Germ-free mice have been exposed to normal housing conditions at different times though their development. Exposure to normal housing conditions has revealed colonization of the sterile GI tract of germ-free mice with normal populations of gut bacteria. This also results in normalization of the undeveloped immune system that is present in germ-free mice. When adult germ-free mice were colonized with normal bacteria, they continued to show reduced anxiety-like behaviors, suggesting that the absence of gut bacteria early in development has a permanent effect on the brain wiring related to anxiety and exploratory behavior.[14, 16] In contrast, when germ-free mice were colonized early in life as pups or during adolescence and then tested in adulthood, normal anxiety-like behavior was observed,[13, 15] suggesting that microbiota influence the way the brain is wired early in development.[17]

In addition to studying mice, researchers have used antibiotic treatment to manipulate gut bacteria. Exposure to antibiotics in drinking water has been shown to lead to reduced numbers of gut bacteria in mice and to a reduction in the diversity of the bacterial population.[18] Consistent with work in germ-free mice, mice exposed to antibiotics for a single week showed increased exploratory behavior and reduced anxiety-like behavior, an observation that was linked to changes in the bacterial profile.[19] Two weeks following the end of the antibiotic treatment, both the bacterial profile and the behavior returned to normal, suggesting that transient changes in gut microbiota can influence behavior.[19]

On the Right Paths

Having established connections among gut bacteria, the brain, and behavior, it is intriguing to consider the ways that microbiota may communicate with the brain. Certainly, classical thinking tells us that there are neural connections from the body to the brain through peripheral nerves, and, in particular, the vagus nerve, which provides information from the gut to the brain. Evidence that bacteria in the GI tract can activate the vagus connection to the brain comes from work showing that administering food-borne pathogens, such as *Citrobacter rodentium* and *Campylobacter jejuni,* to mice activated vagal pathways and related brain regions.[20, 21] This neural activation occurred in the absence of a peripheral immune response, suggesting the presence of a direct link between the bacteria in the gut and the nervous system. In a recent study, feeding healthy mice probiotics, or "good bacteria," decreased anxiety-like and depressive-like behaviors compared to control mice,[22] while a related study showed that feeding mice probiotics activates neurons in the hypothalamus, a brain region known to play a role in stress reactivity.[23] In the latter study, the activation of neurons in the hypothalamus was greater when mice were fed infectious bacteria, leading to a robust peripheral immune response. This suggests that both peripheral nerves and blood-borne immune-signaling molecules can contribute to gut-brain communication.[23] At the level of the hypothalamus, the brain's autonomic nervous system control center, there is considerable evidence that psychological, physiological, and pathological challenges can activate the hypothalamus and turn on the body's stress response. It is fascinating that the communication pathways from gut microbiota to the brain also lead to activation of this key brain region.

While the above-noted work establishes a neural pathway from the gut to the brain, a second important pathway for communication is the immune system. The immune system has two components, the innate immune system and the adaptive immune system. In germ-free mice, the adaptive immune system is undeveloped. Since gut microbiota are essential for immune system development, germ-free mice can be considered to have a low level of inflammation. When we consider the link between inflammation and anxiety-like behavior, we observe that a low inflammatory state is associated with low anxiety-like behavior, whereas higher levels of inflammation lead to increased anxiety-like behavior [17] For example, infection with the parasite *Trichuris muris* in mice results in gut inflammation and increased anxiety-like behavior.[24] In addition, chemically induced gut inflammation in an animal model of colitis also results in gut inflammation and increased anxiety-like behavior.[24] Evidence that the microbiota acts as a modulator of this immune-related increase in anxiety-like behavior is provided in the same reports stating that treatment with probiotic *Bifidobacterium longum* alleviated the anxiety-like behavior.[24,25] These observations suggest that probiotic treatment may have the potential for treatment of inflammation and related anxiety symptoms.

Commensal bacteria play an important role in maintaining the integrity of the intestinal tract, and in situations of stress or disease, increased intestinal permeability can contribute to increased inflammation.[26, 27] Increased intestinal permeability, sometimes referred to as "leaky gut," can lead to translocation of bacteria out of the lumen of the GI tract across the intestinal layer. This is an additional pathway that microbiota and pathogenic bacteria use to communicate with the brain via the immune system or through activation of local neurons in the enteric nervous system (ENS). The ENS is a part of the autonomic nervous system that is housed in the gut and is responsible for gut motility and other normal gut functions.[28] It is a vast network of neurons that are the first points of contact for microbiota in the intestinal lumen and are an important component of the brain-gut axis.

The Stress Factor

One of the most common clinical features of depression is dysregulation of the stress response system, the hypothalamic-pituitary-adrenal (HPA) axis.[29] As previously noted, in response to psychological, physiological, and pathological challenges, neurons in the hypothalamus are activated and signal the pituitary to release adrenocorticotrophic hormone into the bloodstream, which in turn activates the adrenal gland to release the stress hormone cortisol. The stress response, or HPA activation, is part of our normal homeostatic processes, and yet, in depression, it is often overactive or, in some cases, underactive.[29] One of the first studies considering stress and microbiota demonstrated that germ-free mice have an overactive stress response.[23] A more recent study has shown that stress exposure during early life in rats disrupts the microbiota profile and leads to increased stress reactivity in adulthood.[30] Importantly, in this study, treatment of rat pups with probiotic *Lactobacillus sp.* normalized stress hormone levels.[30] Early-life stress also leads to increased depressive-like behavior in adult rats, and a similar study showed that treatment of rats exposed to stress during early life with the probiotic *Bifiodo infantis* reduced the depressive-like symptoms in adulthood.[31] Overall these recent studies imply a link among microbiota imbalance, stress-related behaviors, and stress reactivity, and also suggest that probiotic treatment may be a good approach to treating stress-related symptoms.

To date, researchers have done little work related to stress and microbiota in humans, and in particular, there have been no studies that directly link microbiota to depression or anxiety. The most promising of the clinically related work shows that probiotic administration in people may have antidepressant or anxiety-reducing effects. In one study, healthy subjects were given probiotics for 30 days. Researchers used several questionnaires to test the effects of probiotics on stress, anxiety, depression, and coping. Their results showed that the probiotics group had less psychological stress than the control group did.[32] In a separate study, researchers were able to show that healthy people with low mood at the beginning of the study showed improvement in mood following probiotic administration for three weeks.[33] Finally, in a clinical study on individuals with chronic fatigue syndrome, administration of probiotics over a two-month trial resulted in fewer anxiety-related symptoms.[34] These studies show that clinical trials to date support a role for microbiota in anxiety and depression, and also demonstrate the potential for treatment with probiotics.

Moving Forward

There is no doubt that research in the last decade has established a link between gut microbiota and brain function in mice. We have learned several things: a. gut microbiota are a large population that is important to healthy metabolism and brain function, b. gut-brain communication pathways include neural connections, c. gut microbiota are significant during early development and can influence the wiring of stress circuitry in the brain, and d. probiotics, or "good bacteria," have been shown in animal and human studies to have a beneficial impact on mood symptoms. While these findings are both exciting and promising, we should not make the mistake of thinking that we have found the answers to all clinical situations related to mood. Microbiota—certainly an important modulator of health—must be considered as one component of a complex, multifaceted communication system needed to establish a healthy balance for brain development and function. The research is flourishing across the world as scientists strive to learn more. Stay tuned.

8

Do Cytokines Really Sing the Blues?

By Charles L. Raison, M.D., and Andrew H. Miller, M.D.

Charles Raison, M.D., is associate professor in the Department of Psychiatry, College of Medicine, and the Barry and Janet Lang Associate Professor of Integrative Mental Health at the Norton School of Family and Consumer Sciences, College of Agriculture and Life Sciences, University of Arizona. Raison examines novel mechanisms involved in the development and treatment of major depression and other stress-related emotional and physical conditions. The recipient of several teaching awards, Raison has received research funding from the National Institute of Mental Health, the National Center for Complementary and Alternative Medicine and the Centers for Disease Control and Prevention. Raison is also a mental health expert for CNN.com.

Andrew H. Miller, M.D., the William P. Timmie Professor of Psychiatry and Behavioral Sciences at Emory University School of Medicine in Atlanta, studies brain-immune interactions as they relate to stress and depression. His work focuses on the mechanisms by which cytokines cause depression in humans and nonhuman primates using chronic administration of the innate immune cytokine, interferon-alpha as a model of chronic immune stimulation. Miller has also studied the impact of cytokines on neuroendocrine regulation as well as sleep. Miller and his group have conducted clinical trials examining the efficacy of cytokine antagonists in patients with treatment-resistant depression. In addition to his research work, Miller is the director of Psychiatric Oncology and co-leader of the Cancer Prevention and Control Program at Emory's Winship Cancer Institute.

The World Health Organization predicts that depression, which already affects about 10 percent of the population in the United States, will be the world's leading medical condition by 2030. Evidence suggests several causes for depression, including traumatic life events, disease, poison, and nutritional deficiencies. Many of these causes are associated with elevated levels of inflammatory biomarkers in the blood, which may in turn lead to inflammatory changes in the brain. Our authors examine what the latest research reveals about the link between inflammation in the brain and depression, and how a better understanding of that link can play a critical first step in the personalization of care.

IT IS QUITE THE RAGE THESE DAYS to say that depression is an inflammatory disorder. This conclusion is based on hundreds of studies showing that groups of people with depression have elevated levels of a variety of peripheral inflammatory biomarkers, especially two cytokines—tumor necrosis factor (TNF) and interleukin (IL)-6—and an acute phase reactant called C-reactive protein (CRP).[1] To a large degree, those of us who study the role of inflammation in depression are responsible for spreading the idea. But with the passage of time and the ongoing accumulation of findings from a variety of scientific disciplines, it is becoming increasingly apparent that this characterization, while attractive in the popular press, confuses as much as it clarifies, and is in deep need of refinement.

Suppose a patient comes to see the doctor with a painful, swollen right knee. The doctor suspects rheumatoid arthritis (RA), a classic inflammatory condition, and sends off an aspirate of synovial fluid from the offending knee for analysis. Given that concentrations of inflammatory cytokines such as IL-6 are typically many times higher in the synovial fluid of an affected RA joint than in the blood of a healthy adult, what would be the doctor's conclusion if the joint fluid came back with normal cytokine levels? Almost certainly RA would be ruled out as the cause. No inflammation, no inflammatory disorder.

Now consider an analogous situation, but this time the patient complains of depression. The doctor orders a blood test to measure levels of inflammatory biomarkers. The results come back, and the doctor has a dilemma. The patient is weeping and utterly downcast; shows signs of apathy; can't sleep, eat, or concentrate; and has suicidal thoughts. But all the patient's inflammatory measures are completely normal. Does the doctor conclude that because depression is an inflammatory condition, the patient isn't depressed, no matter how miserable the patient feels? Conversely, if the doctor orders a blood test on someone in good spirits and the test comes back with evidence of increased inflammation, should the doctor insist that the patient is depressed, no matter how good the patient feels?

What these clinical vignettes illustrate is that increased inflammation—however measured—is neither necessary nor sufficient for the diagnosis of depression, and that, therefore, depression cannot be considered an inflammatory disorder without doing a fair amount of damage to the notion of what it means to be an inflammatory disorder. So how should we best understand the association between inflammation and depression? We believe the best available data suggest the following: Depression is not an inflammatory disorder per se; rather, in some patients inflammatory processes appear to contribute significantly to the development and maintenance of symptoms of depression. This perspective unifies much of the scientific literature in the field, while simultaneously suggesting novel diagnostic and treatment approaches for the subgroup of depressed patients with increased inflammation.

The Unspoken Truth

A headline such as "Depression Is Associated with Increased Inflammation" reflects a failure to consider that studies generally indicate that the mean value for a given inflammatory biomarker tends to be higher in depressed groups than in groups of nondepressed individuals. This does not mean that all subjects with depression have higher inflammatory levels than all nondepressed comparison subjects, or that the inflammatory levels in depressed patients are abnormal. Indeed, the mean differences in inflammation between depressed and nondepressed individuals are modest, and mean values for inflammatory markers in groups of depressed individuals are typically in the "normal" range when such norms are established, as is the case with C-reactive protein.

By way of illustrating this point, levels of C-reactive protein in people with autoimmune and inflammatory conditions or acute infection range from 10 to over 100 mg/L,[2, 3, 4] versus depression or no diagnosis, where C-reactive protein ranges from 0 to 10 (with a CRP <3 mg/L generally considered the upper limit of normal).[5] Figure 1 below illustrates the other truth about inflammation and depression that is not always adequately emphasized. Despite differences in mean levels of C-reactive protein or IL-6 in this case, there is a huge overlap between people with depression and those without, and in any given study, the highest inflammatory value might be found in a control subject, while the lowest might be found in the depressed group.[6]

The first point—that the inflammatory activation observed in people

Figure 1. Plasma Interleukin-6 (IL-6) Concentrations of Healthy Comparison Subjects, Comprison Subjects With Major Depression, and Cancer Patients With and Without Major Depression

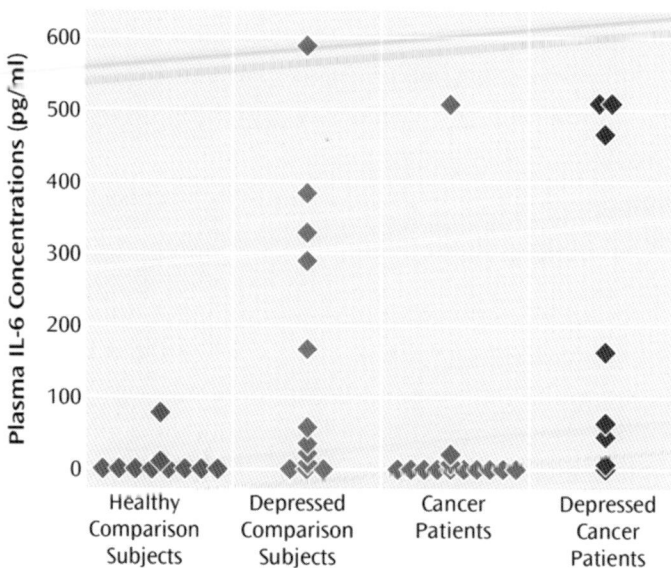

with depression is modest—might tempt us to dismiss the potential relevance of inflammation to the pathophysiology of depression. But this would be both a serious mistake and a profound misunderstanding of the huge effect that small physiological differences can have over time if they are consistently skewed in one direction. As it turns out, depression is far from being alone as a condition characterized by reliable—but often only mildly increased—inflammatory activity. Other modern illnesses with evidence of moderately increased inflammatory signaling include cardiovascular disease, stroke, cancer, diabetes, and dementia. Conversely, even minor increases in inflammation—such as the ones observed in depression—are enough to strongly predict the development over time of many of these modern disease states, including depression.[5,7]

The second point—that there is a high level of overlap of inflammatory biomarker levels between depressed and nondepressed groups—raises a more complex issue. When we say that groups of depressed people tend to have elevated levels of inflammatory biomarkers, what we really mean is that within any depressed group, there are individuals with levels that are significantly higher than those seen in the vast majority of healthy, nondepressed people, whereas there are many other depressed people with perfectly normal values. It is a dirty little secret of sorts that the one-third or so of depressed individuals with elevated inflammation have been pulling all their noninflamed, depressed colleagues along with them in publication after publication, giving the world a slightly misguided sense that depression—as a whole—is driven by increased inflammation.

The critical question is whether inflammation is relevant to depression as a whole or only to individuals with chronically elevated inflammatory biomarkers. And if depression is relevant only to those with increased inflammation, how much of an increase needs to exist before it reliably contributes to depressive pathogenesis? And might it be the case that people with depression and low levels of inflammation are just more sensitive to the depressogenic effects of inflammatory activity, so that even low levels disrupt brain functioning in ways that promote the disorder?

What Cytokine Antagonism Has Taught Us

A recent study from our group provides some surprising, tentative answers to these questions. Determined to see if peripheral inflammatory processes really contribute to depressive pathogenesis, we decided to put the theory to the test by examining whether blocking the inflammatory cascade would eradicate depression in patients who were otherwise medically stable. To test this as rigorously as possible, we elected to use a medication called infliximab, which is not believed to cross the blood-brain barrier—a tight layer of cells and tissue that separates the brain from the rest of the body— and has no biological effects other than to potently block the activity of TNF, the cytokine that along with IL-1beta is most responsible for initiating the inflammatory response.[8] We measured pretreatment levels of peripheral inflammation in 60 patients with treatment-resistant depression, which has been shown to have a special relationship with increased inflammation, in part related to the ability of cytokines to sabotage and circumvent the mechanism of action of antidepressants. Patients were then randomized to receive three infusions of either infliximab or saline in a blinded manner over a six-week period. We followed depressive symptoms during this period and for six weeks following the final infusion.

The results were unequivocal. For the group as a whole, infliximab was no better than placebo—in fact, it was nearly identical. This strongly suggests that inflammatory processes are not directly causal for the entire spectrum of pathology that we currently characterize as depression. Interestingly, however, we found a significant relationship between pretreatment levels of inflammation and clinical response. Indeed, for people with baseline levels of C-reactive protein above 5 mg/L, infliximab outperformed placebo to the same degree that standard antidepressants do.

What we were also not expecting was that for depressed individuals with low levels of peripheral inflammation, placebo far outperformed infliximab, suggesting that blocking inflammation was not just neutral, but was actually doing something profoundly negative in these individuals. At the least it was blocking people's ability to respond to placebo. More worrisome is the possibility that it was blocking potentially positive effects of inflammation. Perhaps we should not have been as surprised as we were. Animal work by psychobiologist Raz Yirmiya and others had already suggested that at lower concentrations, inflammatory molecules such as TNF play important roles in processes associated with positive brain functioning, including neurogenesis, synaptic scaling, and long-term potentiation.[9] More recently, the antidepressant-like effect of deep brain stimulation (DBS) in rodents has been shown to be associated not with electrode placement or even with the passage of electric current, but rather with the local brain inflammation induced by needle placement.[10] Consistent with this, the use of anti-inflammatory agents in humans seems to block the efficacy of DBS in patients with treatment-resistant depression. This negative association between anti-inflammatory agents and response was also noted with standard antidepressants in the STAR*D trial.[11]

Taken together, these results suggest that the relationship between inflammation and depression may be U-shaped, with both very low levels and high levels of activity posing risks, albeit for different reasons. Nevertheless, the modern world is so replete with depression risk factors that also promote inflammation—including obesity, sedentary lifestyle, sleep loss, psychosocial stress, processed foods, air pollution, and perhaps hygienic practices (separating us from exposure to a range of coevolved microorganisms and parasites with powerful immunomodulatory and anti-inflammatory effects[12])—that the association between inflammation and depression is hardly surprising. This is especially true given that relatively mild elevations in peripheral inflammatory activity—when chronic—seem sufficient to set in motion physiological mechanisms that are depressogenic, particularly in individuals made vulnerable by early environment and/or genetic inheritance.[13] It is to these mechanisms that we now turn.

How Inflammation Produces Depression

When we started exploring how inflammation produces depression, we were convinced that depressogenic pathways unique to the immune system would be identified, leading to the conclusion that depression that occurs in response to stress or comes "out of the blue" is biologically different from depression induced by immune activation. We couldn't have been more wrong.

One of the most surprising, and consistent, findings over the last decade has been that inflammatory processes induce depression because they are capable of tapping in to every known risk pathway. In this regard, inflammation functions like psychological stress, which has similar wide-ranging, and generally depressogenic, biological effects. In fact, inflammation causes most of the same changes in the brain and body that have been repeatedly observed in animals and humans exposed to stress, especially when the stress is chronic and of a psychosocial nature.1

One of the best models for understanding the impact of chronic inflammation on humans has been treatment with the cytokine interferon (IFN)-alpha for either cancer or hepatitis C virus infection. IFN-alpha causes full-blown depression in a respectable minority of patients, produces depressive symptoms in a far higher percentage, and produces an improvement in mood in no one but the unfortunate few who develop mania in response to the treatment.[14] Our understanding of the central-nervous-system (CNS) effects of peripheral inflammation also have been augmented by studies of humans who became acutely inflamed as a result of receiving a dose of lipopolysaccharide or a typhoid vaccine.

Figure 2 below provides a schematic of the neurobiological pathways known to be influenced by both acute and chronic inflammatory stimuli in humans, ultimately leading to alterations in neurocircuitry and behavior. Neuroimaging studies reliably indicate that peripheral inflammation targets brain regions repeatedly implicated in the pathophysiology of depression, especially the anterior cingulate cortex (ACC) and basal ganglia.[15] Depending on the nature of the stimulus, the impact of cytokines on the ACC is seen most strongly in either the subgenual or the dorsal area.[16,17] Cytokine-induced increases in neural activity in these regions have been associated with the development of mood and anxiety symptoms. On the other hand, peripheral immune activation has been shown repeatedly to impair basal ganglia functioning in ways that are consistent with the known inhibitory effects of cytokines on dopamine signaling in the CNS.[18,19] Reductions in basal ganglia activity have been noted in more posterior regions,

Figure 2. Inflammation
and Depression: Path-
ways to Pathology

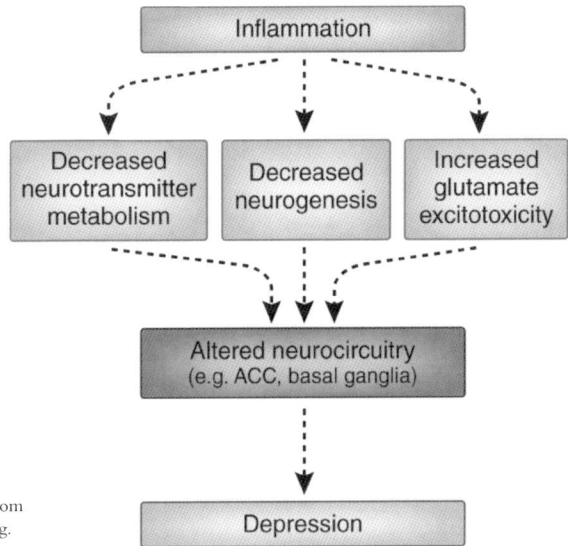

Inflammation

Decreased
neurotransmitter
metabolism

Decreased
neurogenesis

Increased
glutamate
excitotoxicity

Altered neurocircuitry
(e.g. ACC, basal ganglia)

Depression

where they associate with fatigue, and in more ventral regions (such as the nucleus accumbens), where they have been associated with the development of anhedonia, a psychological condition characterized by an inability to experience enjoyment in normally pleasurable acts.[18]

Regarding the mechanisms of the effects of cytokines on these and other brain regions, cytokines have been shown repeatedly to alter neurotransmitter signaling in the CNS in ways that are relevant to the pathophysiology of depression and its treatment. For example, through activation of the intracellular signaling pathway mitogen-activated protein kinase, cytokines can increase the number and function of the reuptake pumps for serotonin, norepinephrine, and dopamine, which in turn can reduce the availability of these neurotransmitters within the synaptic cleft. This is relevant to depression and its treatment given that most currently available antidepressants act by blocking these reuptake pumps to increase neurotransmitter availability in the synapse.

Cytokines have other effects known to impact neurotransmitter availability. Indeed, by activating enzyme indoleamine 2, 3-dioxygenase, cytokines can shunt tryptophan away from the production of serotonin and into the production of kynurenine. Kynurenine is transported to the brain and can be converted by activated microglia (innate cells in the brain) to the neurotoxic metabolite quinolinic acid. The clinical relevance of this process has been shown by the association between cerebrospinal fluid levels of kynurenine and quinolinic acid, and the development of depression during treatment with IFN-alpha.[1] Moreover, increased quinolinic acid has been found in activated microglia in the ACC of suicide victims who were depressed. Quinolinic acid can impact glutamate signaling in ways relevant to depression, including the stimulation of extrasynaptic N-methyl-D-aspartate (NMDA) receptors, which lead to the downregulation of the production of brain-derived neurotrophic factor (BDNF), a potent inducer of neurogenesis. Consistent with this and other activities of inflammatory cytokines, animal and human studies have demonstrated that increased inflammatory cytokines can reduce central levels of BDNF and neurogenesis, leading to depressive-like behavior.[20]

Depression is not just a brain disease. Indeed, many of the depression-related physiological abnormalities identified at the dawn of biological psychiatry involved the body's stress system and especially the hypothalamic-pituitary-adrenal (HPA) axis. As a group, depressed individuals have been repeatedly reported to demonstrate increased circulating cortisol and concomitant glucocorticoid resistance (e.g., decreased sensitivity to the inhibitory effects of glucocorticoids on HPA axis regulation and inflammation).[21] Depressed patients also show a flatter diurnal pattern of cortisol secretion than do healthy control subjects. Strikingly, inflammatory cytokines have been shown to be capable of producing all these abnormalities, including glucocorticoid resistance, and in the context of treatment with IFN-alpha, flattening of the cortisol slope strongly predicts the development of depression.[22]

Do Stress and Inflammation Always Cause Depression?

Further suggesting a link between inflammation and depression is the fact that psychosocial stress—which is a primary risk factor for depression development—reliably activates peripheral inflammatory pathways in humans.

This activation is measured either as increases in plasma concentrations of inflammatory cytokines such as IL-1beta and IL-6 or as increased activation of the intracellular inflammatory transcription element nuclear factor-kappa beta (NFkB).[1]

Although not established in humans, psychological stressors in laboratory animals have been shown repeatedly to increase levels of proinflammatory cytokines in the CNS, especially in areas, such as the hippocampus, that are integrally involved in the mammalian stress response and repeatedly implicated in depression. Importantly, in these animal models, the behavioral and biological effects of stress (such as reductions in BDNF) can be made more tolerable by blocking the stress-induced increases in CNS inflammatory signaling.[20] In animal models, stress has been shown to activate CNS microglial cells, which take on an inflammatory phenotype when activated.[23] Of relevance to depression, several postmortem studies point to evidence of microglial activation in individuals who died by suicide.[24] Also linking inflammation to depression is the fact that individuals at high risk for depression—such as those exposed to early-life adversity—respond to laboratory psychosocial stressors with more robust inflammatory responses than do others.[25] Inflammatory responses to these types of stressors have, in turn, been shown to predict the future development of depression.

The Implications for Care

The science is clear: While depression is not an inflammatory condition, inflammation can cause depression, and inflammatory cytokines are clearly a factor. Although it is simple, this idea has profound implications. It suggests that inflammatory processes will be highly relevant to some individuals with depression, irrelevant to others, and perhaps even of benefit to a minority of depressed individuals. Recent data suggest that something as simple as the widely available blood test for C-reactive protein may identify individuals with greater or lesser likelihood of responding to anti-inflammatory therapeutic strategies and, by extension, individuals for whom inflammation is more or less of a causative factor. Such identification of patients with depression and increased inflammation represents a critical first step to the personalization of care, allowing specific treatments to be directed to specific pathologies that can then be monitored as a function of response. Such a development represents a game changer in psychiatry and truly emphasizes that when it comes to treatment for depression, one size does not fit all.

9

ADHD

10 Years Later

By Philip Shaw, Ph.D.

Philip Shaw, Ph.D., is an Earl Stadtman Investigator at the Neurobehavioral Clinical Research Section of the National Human Genome Research Institute and an adjunct faculty member of the National Institute of Mental Health. His main interest is the genetic and environmental factors that influence the development of brain and behavior. He has degrees in experimental psychology and medicine from Oxford University and a Ph.D. in psychological medicine from the Institute of Psychiatry in London. He completed residencies in internal medicine and psychiatry in England, and is a member of both the Royal College of Physicians and the Royal College of Psychiatrists.

 *Estimates of children struggling with attention-deficit/hyperactivity disorder (ADHD) vary, but the Centers for Disease Control puts the number at a stunningly high 25 percent. Whatever the number, ADHD affects too many children at school, at home, and with their peers, and often persists into adulthood. The cause is as yet unknown, although genetic factors and their interaction with the environment are known to be pivotal. Ten years ago a landmark study showed that the structure of the brains of children with ADHD differs from that of unaffected children. Since that study, enhancements in imaging have given researchers a better look at key hubs in the brain and how they network—advances that could prove useful in the control and treatment of ADHD in both children and adults.*

IMAGINE TWO CHILDREN, both 8 years of age, sitting in science class. Both focus on the chalkboard and the teacher as she explains why Earth orbits the sun. The first child, Toby, is engrossed in the topic; he gazes intently at the teacher and at diagrams of the solar system. The second child, Susan, tries to stay focused but keeps getting distracted, first by the whispering of a classmate, then by the sunny day outside, and finally, by the antics of the classroom hamster. She fidgets constantly and is frequently out of her chair, despite the teacher's reminders to stay seated.

What brain events underpin the different behaviors in these two children? Think first of Susan's fidgeting. This apparently simple movement is not the result of one isolated part of the brain; rather, it's the result of a network of brain structures acting in concert to produce movement. These networks are made up of key components, or hubs, that are connected by specialized links known as white matter tracts. In the planning and execution of movements, the key hubs include parts of the prefrontal cortex (the motor and premotor cortex) and deeper brain structures.[4, 19, 20] The deeper structures include the putamen, the thalamus, and the cerebellum. The putamen, thalamus, and cerebellum add and integrate relevant information, which is then relayed back to the cortex.

Now consider each child's ability to stay focused on the teacher and directed toward the goal of learning. This ability also depends on brain structures that act together. In this case, the network is responsible for controlling a highly complex cognitive act, which it achieves partly by integrating information from the lateral prefrontal and parietal cortices. Often this information is further relayed through deeper structures in order to guide behavior, to make decisions, and to solve problems. Considering all of the brain activity necessary to stay focused in the classroom, it is easy to see how problems with these networks could translate into challenges with attention and motor control.

Problems with the control of attention, impulses, and movement can be severe. In the extreme, they are strongly associated with a child struggling academically at school, sometimes having difficulty forming friendships with peers, and causing problems at home.[8, 11, 30] Such problems can prompt a full clinical assessment, which generally involves talking with parents and teachers. The result may lead to a diagnosis of attention-deficit/hyperactivity disorder (ADHD).

How has imaging of the structure of the brain added to the understanding of problems in the domains of attention, impulse, and motor control—the cardinal features of ADHD? Three points emerge. First, key regions or hubs in the networks mediating the control of attention and action sometimes show structural differences between groups with and without ADHD. Second, the hubs' physical connections, formed by white matter tracts, may also differ in the brains of people with impaired attention. Finally, some of the structural differences associated with ADHD are not fixed and static, but rather change as a child grows.

What Do We Know?

What has brain imaging found? Just over 10 years ago, a landmark study used anatomic MRI to compare 152 children with ADHD to a group of 139 children with no symptoms. The ADHD group showed a slight reduction—about 3 percent—in total brain volume.[5] This did not mean the children with ADHD were less intelligent; they were as intelligent as the comparison children in the study. Rather, it suggested that severe, impairing problems with attention, impulse, and activity control are associated with differences in brain structure.

Since this study, marked advances in acquiring and analyzing brain images have enabled researchers to pinpoint the brain regions most tightly linked with ADHD. These advances allow the identification of structural differences in the hubs of networks that most strongly contribute to challenges with control of attention and movement. Three hubs are shown in the figure below. Many studies have defined brain-volume change in terms of voxels.

When we pool the results of these studies, the aggregate findings point to key hubs that might be important in the control of attention. The first candidate hub lies in the striatum, a deep brain structure made up of two substructures, the caudate and the putamen. The striatum, which lies near the center of each hemisphere, interacts richly with multiple other brain regions. It is a hub in networks supporting many cognitive skills, such as the flexible control of attention and motor planning. The striatum is slightly smaller in groups with ADHD compared to those without. The effect is

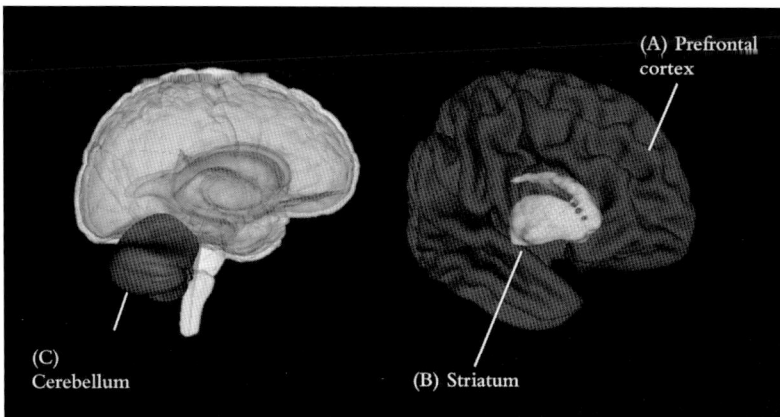

(A) Prefrontal cortex

(C) Cerebellum

(B) Striatum

The key components or "hubs'" of the networks that support the control of action and attention are shown. These include (a) regions of the prefrontal cortex (particularly the dorsolateral cortex, and the midline cingulate cortex, which is not shown); (b) parts of the striatum; and the (c) cerebellum, particularly the midline vermis. These regions are interconnected.

more prominent on the right side of the putamen and the anterior parts of the caudate. Many individuals with ADHD find attention-control and motor-planning skills particularly challenging. It thus makes perfect sense that this region would show structural change. More recent work suggests that changes to the striatum apply not only to its volume, but also to its surface contours, with reports of "indentation" in parts of the striatum in groups with ADHD.[26] At an even deeper level in the striatum, images show changes in the density of receptors for dopamine, which is one of the chemicals in the brain that is important in communication between cells.[29]

A second set of candidate hubs has been localized to the prefrontal cortex. This set emerged partly from studies that mapped both cortical thickness and surface area at thousands of points across the brain. Most of these studies indicate that the dorsolateral prefrontal cortex is slightly thinner in groups with ADHD compared to those without the condition.[1, 2, 17, 23] This frontal region is a central hub in the networks controlling the most complex executive functions, such as allocation of attention, planning, and decision making. A second structurally altered hub lies in the cingulate cortex, which is pivotal in monitoring the environment and adjusting behavior in response to feedback. Tellingly, both of these frontal areas belong to networks that include parts of the striatum.

A child's environment has an exquisite temporal structure; successful navigation requires the ability to process information that changes extremely quickly—on the order of milliseconds. [6, 18, 21, 22, 25, 27] For example, a conversation requires the ability to judge exactly when someone is about to stop speaking and then to respond within an appropriate time period. A child may appear impulsive because they fail to judge exactly the correct amount of time needed before responding to a teacher, for example, and interrupt. The cerebellum is a critical part of the networks responsible for such temporal information processing. Decreased volume of the cerebellum has been found in groups with ADHD, compared to groups without ADHD.[3, 16] The midline section of the cerebellum—called the vermis—has been most strongly linked with attention problems. Notably, the vermis is richly interconnected with the cortical hubs mentioned earlier.

In short, a series of networks controls attention and action. Within these networks, three hubs emerge as important: centers in the prefrontal cortex, the striatum, and the cerebellum. This concept of networks provides hope for overcoming attention problems. For example, one hub might be able to compensate for another, a point we will return to later.

MRI as a Game Changer

Magnetic resonance imaging (MRI) allows us to look at the brain. It leverages the fact that different tissues and regions can produce a different "signal" when placed in a strong magnetic field. This approach is safe, as it does not use ionizing radiation. Just two of its many applications are considered in this article. The first is the physical definition of the brain's various structures. Scientists once established these boundaries by tracing around entire structures with their hands, but over time they have increasingly used objective computational approaches to define structures at a much finer level of resolution. For example, in one approach, scientists divide the brain into hundreds of thousands of tiny boxes, or voxels (mentioned earlier), and measure the volume of each voxel. In another, they look at the thickness of the cortex at hundreds of thousands of points across the brain. Both approaches afford exquisite precision in mapping anatomical differences.

The second relevant application of MRI helps scientists define the brain's "wiring." White matter tracts, which often connect different brain regions, have physical properties that can be measured by MRI. Water molecules in the brain move randomly unless they are constrained by a physical barrier. White matter tracts, or, more precisely, the myelin sheaths that form tracts, are one such barrier. These tracts can be mapped using a form of MRI called diffusion tensor imaging (DTI), which enables us to better understand the connections between brain regions.

When describing the findings, it is important to stress that the brain-structure differences emerge from studies that compare one group with ADHD against a group without it. On average, the size of certain brain structures differs between these groups. However, the size of brain structures varies greatly within both groups, and the size of brain structures also overlaps significantly between the groups. For example, the volume of a structure called the striatum is lower, on average, in groups of people who have ADHD than it is in groups of people who do not have ADHD. This means that the distribution (or the spread) of striatal volumes in ADHD is shifted, compared to the distribution of areas in the comparison or control group. There is, however, still much overlap between the distributions. In other words, the volume of an individual's striatum is insufficient information for a diagnosis; the diagnosis is still based on a careful assessment that includes feedback from family members and teachers. Nonetheless, the robust group differences found in neuroimaging studies give us invaluable insights into the neurobiology of inattention, impulsivity, and hyperactivity. Researchers are benefiting from the emerging picture of group-level structural change in brain regions that are pivotal in the cognitive processes most affected in ADHD. Such work not only guides research, but also may inform future treatment.

The Brain's Wiring

A good network requires good communication. Carrying information quickly and efficiently between hubs in a network is a prerequisite for optimal brain functioning. Using techniques such as DTI, we can map the brain's communication pathways. Scientists debate about the exact physical properties that DTI captures, but broadly speaking, one measure—the fractional anisotropy—reflects the structural integrity and organization of white matter tracts. Some themes emerge when we pool the results of DTI studies comparing groups of individuals with and without ADHD.[13, 28] First, fractional anisotropy is lower in some regions of the brain in groups with ADHD compared to groups without the condition. These changes are prominent, but they are not confined to the tracts that link the hubs mentioned earlier. To give a specific example, reduction in fractional anisotropy localizes to a region near the tracts that connect the cerebellum with prefrontal cortical hubs important for motor control. Another region that shows change contains the white matter tracts that connect different cortical hubs (the superior longitudinal fasciculus).

Accelerating the development of new methods to map brain

structural connectivity is the Human Connectome Project, a five-year project sponsored by 16 components of the National Institutes of Health and split between two consortia of research institutions. The project is the first large-scale attempt to collect and share data of a scope and level of detail sufficient to begin the process of addressing fundamental questions about human connectional anatomy and variation, with the hope of gaining a richer understanding of the connectivity problems that underpin challenges in attention and impulse control.

Cause or Effect?

Are these brain structural changes the result, rather than the cause, of problems with inattention, impulsivity, and hyperactivity? Evidence from studies on families indicates that brain anatomic changes are not merely the result of having symptoms of ADHD, but rather they are a causative factor. In these studies, siblings with and without ADHD are compared. The studies indicate that changes in brain structures in unaffected siblings that resemble the structural changes in siblings with ADHD cannot be due to ADHD symptoms, as the unaffected siblings have no symptoms. Rather, the brain changes reflect familial, likely genetic, influences on brain anatomy. One study demonstrated that children with ADHD and their unaffected siblings share similar structural brain differences, including a slight reduction in the volume of the prefrontal cortex.[9] Another study found white matter tract structure changes in both individuals with ADHD and siblings without ADHD.[15]

Similar family studies have long been used to isolate genes that contribute to mental-health problems. An exciting future direction is to adopt this approach in neuroimaging by identifying the brain changes that track attention problems within families. Such work is particularly compelling given the possibility of sequencing the genome—that is, spelling out the entire genetic code of an individual. Harnessing the power of both genomic sequencing and rapidly advancing neuroimaging techniques will yield great insights into the mechanisms underpinning the control of attention and action and why they sometimes go awry.

The Dynamic, Developing Brain

Think back to Susan, the child whose problem with staying focused was impacting her ability to learn. How do children with problems similar to Susan's fare as they grow up? Several large studies follow children with a diagnosis of ADHD into adulthood.[24] Full-blown ADHD persists around 20 percent of the time. While a further 50 percent have symptoms that impair day-to-day living, they do not have enough symptoms to warrant a formal diagnosis. The remainder shows more robust improvement, with resolution of most symptoms. Generally, hyperactivity and impulsivity tend to improve more than inattention does. As with many challenges of childhood, ADHD has a highly variable course. Understanding the brain changes or trajectories that underpin these variables could help us devise treatments to keep brain development on track.

Longitudinal data can be particularly powerful in defining these trajectories. In one study, children with ADHD underwent repeated assessment of their symptoms along with MRIs as they grew into adulthood.[24] This approach allowed researchers to link the trajectories of brain development with adult outcomes. Some differences in trajectories emerged in a prefrontal cortical region mentioned earlier: the cingulate cortex. This region is a key hub in a network that keeps an individual on task and monitors the environment for relevant information.[19] In the study, individuals whose attention improved started with a thinner cortex in childhood, but the cortex eventually grew to the same thickness as a comparison group that never had ADHD. By contrast, those who had persistent inattention into adulthood did not show this convergence; rather, their cortices showed a fixed difference. There was a great deal of variability in these trajectories, but overall differences in the development of an "attention hub" were associated with later outcome. Movies provide more information than snapshots do: Likewise, developmental differences important for the control of action and attention can often be captured better by longitudinal than cross-sectional data.[14]

Developmental Perspective

Thinking of the brain as a developing network might inspire new approaches to help people overcome impairing problems with attention and impulsivity. Perhaps intact hubs in a network can be recruited to compensate for suboptimal performance of other networks.[7] Or perhaps through techniques such as cognitive-skill training, people with ADHD can improve communication between key hubs.[12] Our emphasis on the importance of taking a developmental perspective underscores the brain's capacity for beneficial change. It is exciting to think how the enormous potential for brain plasticity in childhood and adulthood could be harnessed to improve attention skills and to boost motor and cognitive control.

10

Lewy Body Dementia

The Under-Recognized but Common Foe

*By James E. Galvin, M.D., M.P.H., and
Meera Balasubramaniam, M.D.*

James E. Galvin, M.D., M.P.H., is a professor of neurology, psychiatry, nursing, nutrition, and population health at New York University Langone School of Medicine. He serves as director of the Pearl I. Barlow Center for Memory Evaluation and Treatment and also as director of the Lewy Body Dementia Research Program. Dr. Galvin received his bachelor of arts degree in chemistry from New York University in 1986, his master of science in nutrition from Rutgers University in 1988, his M.D. from the University of Medicine and Dentistry of New Jersey in 1992, and his master of public health from St. Louis University in 2004. He has authored more than 115 scientific publications covering basic, clinical, and translational science in the area of neurodegenerative disorders, dementia, and cognitive aging.

Meera Balasubramaniam, M.D., M.P.H., is currently a clinical fellow in geriatric psychiatry at the New York University Langone School of Medicine. After obtaining her medical degree at Seth G.S. Medical College and KEM Hospital in Mumbai, India, she trained in epidemiology at the University of Texas at Houston School of Public Health and subsequently completed her residency in psychiatry at Duke University. Her clinical and research interests include delirium, behavioral management of dementia, and late-life mood disorders. She is a member-in-training of the American Association of Geriatric Psychiatry and was awarded the Honors Scholarship in 2012.

After Alzheimer's disease, Lewy body dementia (LBD) is the most prevalent progressive dementia of the many cognitive disorders wreaking unspeakable havoc on millions of lives. LBD is characterized by the presence of Lewy bodies, which are abnormal aggregates of a protein called alpha-synuclein, and are found in regions of the brain that regulate behavior, memory, movement, and personality. Many of the symptoms of Alzheimer's, Parkinson's, and LBD overlap, but LBD is more difficult to diagnose. Underdiagnosis is just part of the reason why LBD is unknown to the public and many health-care providers, and why funding for research lags far behind that for almost every other cognitive disorder.

"HE'S NO LONGER THE SAME, nor the person he used to be," says Mrs. A of her husband of 52 years. Mr. A, a 70-year-old retired professor, looks strangely unaffected as Mrs. A helps steady his way as he cautiously approaches the clinic's examination table. Until two years ago, his family saw him as perfectly healthy, a talented musician and the doting father of two adult children. But when he started relying on "sticky notes" to keep up with his daily activities and needed help to handle routine tasks, they began to suspect something was wrong. Over time, he became distant and withdrawn and would stare into space, a pale shadow of his once gregarious self.

Mrs. A shows the neurologist the bruise Mr. A sustained the previous night, following another of his terrible nightmares. Mr. A falls asleep during a portion of the interview, as he often does during the day. Mrs. A says there are "good days and bad days" as she recounts the first time her husband talked about seeing "little animals" that she knew did not exist. While he is unable to recall what he had for dinner the previous night, he remembers "fried rice" when his wife reminds him that they enjoyed his daughter's favorite recipe. As Mrs. A fights back tears, she asks the neurologist, "They say it's Lewy body dementia. Is that the same as Alzheimer's?"

The Recognition Factor

Lewy body dementia (LBD) is not rare. According to the Lewy Body Dementia Association (www.LBDA.org), LBD affects approximately 1.3 million individuals in the United States and is the second most common form of dementia, after Alzheimer's disease (which affects about 5 million).[1,2] LBD actually encompasses two disorders: dementia with Lewy bodies (DLB) and Parkinson's disease dementia (PDD); both represent a spectrum of symptoms involving cognition, movement, behavior, and sleep. The difference in diagnosis largely depends on the order of symptom presentation. If the movement disorder begins more than one year before any cognitive symptom, it is usually referred to as PDD, while any other pattern of symptom presentation is usually referred to as DLB. Both are generally referred to as LBD, even in scientific papers.

It is important to first clearly define what is meant by the term "dementia" and how LBD differs from Alzheimer's. "Dementia" is a general term that describes a progressive decline in cognitive function that represents: (a) a change from previous abilities; (b) interference with everyday functioning; and (c) a condition not caused by another illness. There are many forms of dementia (more than 100), with Alzheimer's considered the most common. While all individuals with dementia experience decline in memory, thinking, language, problem solving, judgment, and the ability to function independently, as well as changes in behavior, there is wide clinical overlap between the different types of dementia. As the diseases progress, the distinctions blur even further. What may make the distinction even more difficult and confusing is that many cases of dementia are actually "mixed," or due to multiple causes. In the case of LBD, nearly 80 percent of individuals will also have brain changes consistent with Alzheimer's, while almost 40 percent of those diagnosed as having Alzheimer's are found to have features of LBD.[3,4] This overlap probably contributes to the difficulty in making a clinical diagnosis of LBD and helps to explain why so many patients and caregivers find the diagnostic experience so frustrating.[5] Not only is there overlap between the separate and distinct LBD and Alzheimer's, but awareness and understanding of LBD are generally lower among both the public and the medical community.

The resulting underdiagnosis of LBD has far-reaching implications. In addition to delaying appropriate treatments to alleviate symptoms, the underdiagnosis may expose patients to potentially dangerous adverse reactions to certain medications (such as classic neuroleptic medications like haloperidol). Early and accurate diagnosis helps families prepare for their role in caregiving, specifically the behavioral management and their own emotional preparation in anticipation of the disease course, one that may have its own unique challenges and burdens.[6,7] The attention to and advancement in research directed at the diagnosis and treatment of LBD significantly lag behind those of Alzheimer's, another consequence of the limited awareness and underdiagnosis.

Drawing Distinctions

Lewy body dementia and Alzheimer's share clinical as well as pathological features, making their separation challenging for even the most seasoned clinician. The presence of mixed forms of the illnesses complicates the picture further. A review of the features of LBD clarifies the picture.[2] The LBD diagnosis is made when a patient presents cognitive decline (i.e., a dementia) with at least two of the following features: (a) Parkinson's-like movement changes, including slowness, stiffness, tremor, or balance problems; (b) visual hallucinations, often of small people, children, or animals; (c) spontaneous changes in alertness, concentration, and attention, called cognitive fluctuations; and (d) a sleep disorder causing people to "act out" their dreams, called rapid eye movement sleep behavior disorder (RBD). These symptoms are fairly specific to LBD, and are not present in most cases of Alzheimer's or other dementias. The challenge for clinicians is to figure out how best to detect these symptoms. In particular, cognitive fluctuations are the most difficult symptoms for the clinician to elicit, unless specific questions are asked, such as those found in the Mayo Clinic Fluctuations Questionnaire.[8]

One clue to proper diagnosis is related to gender. LBD is found more commonly in men, whereas Alzheimer's has a slight preponderance among females. A research study focusing on caregivers' reports revealed that LBD's most common presenting symptoms are memory impairment (57 percent), visual hallucinations (44 percent), depression (34 percent), difficulties with problem solving (33 percent), difficulty with gait (28 percent), and tremor/ stiffness (25 percent). In contrast, almost all (99 percent) individuals with Alzheimer's reported memory impairment as a presenting symptom.[9] While difficulty with memory is a common presenting complaint in both forms of dementia, the nature of the impairment differs, particularly at the onset. Alzheimer's affects the ability to encode new experiences into one's long-term memory, whereas the disorder in LBD can be one that affects retrieval of memory.[10,11] LBD patients may perform worse on visual-spatial tests than Alzheimer's patients, while Alzheimer's patients may perform worse on tests of language function.[12] Neuropsychological testing can help to tease out these differences.

In its early stages, LBD is also more likely to be associated with psychiatric symptoms. The patient not only sees nonexistent people, animals, body parts, or even vehicles, but may describe them in detail. He or she may even respond by talking to the hallucinations.[7] Paranoia toward caregivers and unshakable false beliefs, such as that family members are being replaced by impostors, are more prevalent among individuals with LBD than among Alzheimer's patients (this is called Capgras syndrome).[13] Patients with Alzheimer's frequently develop psychotic symptoms later in the course of the disease, such that the late stages of LBD and Alzheimer's may be indistinguishable.[14]

Some evidence suggests that the ability to plan and organize complex tasks, known as "executive functioning," is affected earlier in LBD than in other forms of dementia, although this is not a consistent finding. Patients with LBD are no longer able to perform routine tasks,[15] and are often found to be withdrawn earlier than those affected by Alzheimer's. This, coupled with partial awareness of their diminishing cognitive abilities, may be associated with higher frequency of depression, apathy, and social withdrawal.[16]

LBD also impairs the functioning of the autonomic nervous system, which regulates body functions such as blood pressure, heart rate, sweating, digestion, and bowel and bladder emptying. The clinical symptom of most concern is a decrease in blood pressure upon standing from a horizontal position, which increases the risk of dizziness and falls in about 15 percent of individuals with LBD. Patients are also found to have constipation (which may even pre-date the onset of cognitive symptoms), diarrhea, urinary dysfunction, excess salivation, decreased sweating, heat intolerance, impotence, and erectile dysfunction.[12,17] Predictably, these symptoms add multiple layers to the care of an individual with LBD, thereby necessitating early and accurate diagnosis.

Neurological Causes

LBD is characterized by the presence of Lewy bodies in regions of the brain that regulate behavior, memory, movement, and personality. Lewy bodies are composed of abnormal aggregates of a protein called alpha-synuclein.[2] Autopsy findings of individuals with LBD have demonstrated amyloid plaques, which are the key pathological findings in Alzheimer's. However, the other characteristic feature in Alzheimer's, the "neurofibrillary tangles," consisting of aggregates of modified tau protein, are typically not associated with LBD.[2,18] Scientists are just beginning to understand the steps of protein aggregation in Alzheimer's and LBD.

The knowledge of changes associated with Alzheimer's is far advanced compared with that of LBD. There are several reasons for this. First, the abnormal protein in LBD, alpha-synuclein, was discovered a decade later than the proteins involved in Alzheimer's. Second, the amount of overall pathology associated with alpha-synuclein occurs at lower levels than amyloid and tau proteins in Alzheimer's, which makes it harder to develop lab tests that measure it. Third, there are fewer genetic causes of LBD that can be used to create experimental models than there are of Alzheimer's. Finally, it has been more difficult to develop animal models of LBD that recapitulate both the pathology and the symptoms of the disease.

The brain uses chemicals called neurotransmitters to send messages from neuron to neuron, and multiple neurotransmitter systems are involved in LBD. A consistent finding has been the dysfunction of the cholinergic system. With LBD, it occurs earlier and to a greater degree than in Alzheimer's.[19] It may also be responsible for LBD's earlier onset of hallucinations and fluctuations.[17] It is noteworthy that one of the main groups of medications currently available to delay the progression of the dementia reverses the deficit in the cholinergic system. The dopamine system is also affected, leading to LBD's movement disorder, one that can be treated with the same medications used to treat Parkinson's.[17] Because there does not seem to be an LBD-specific neurotransmitter, it is likely that the many research programs for Alzheimer's and Parkinson's will also provide potential therapies for LBD.

On brain imaging studies, both Alzheimer's and LBD demonstrate neuronal cell loss and shrinkage (atrophy) of the brain. However, volumetric MRI studies of individuals in similar stages of Alzheimer's and LBD reveal that the specific regions of the brain that control memory—namely the medial temporal lobes—are affected early in Alzheimer's but may be preserved until later stages in LBD.[20] SPECT scan, a type of nuclear medicine imaging technique using gamma rays, has shown that a transporter for the neurochemical dopamine is affected to a greater degree in LBD when compared to Alzheimer's.[21,22] This type of imaging (called a DaTScan) has provided the first diagnostic test for LBD approved by the European Union, but it has not yet been approved by the Food and Drug Administration (FDA).

The Many Challenges

With the myriad symptoms that Mr. A experienced—memory impairment, problem solving and function impairment, fluctuation in cognition, vivid nightmares (RBD), visual hallucinations, and possible postural hypotension—diagnosing LBD correctly can be akin to solving a jigsaw puzzle. Primary-care providers rarely get it right. In a survey of more than 900 LBD patients and caregivers, 50 percent saw three doctors for more than 10 visits over the course of one year before an LBD diagnosis was established.[5] First diagnoses given to the patients were most commonly Parkinson's or other movement disorders (39 percent), Alzheimer's or other cognitive disorders (36 percent), and mental illness (24 percent). Neurologists diagnosed most cases (62 percent), while primary-care providers diagnosed only 6 percent of cases.[5] The remainder of cases (32 percent) are diagnosed by other specialties.

One could only imagine the challenges facing Mrs. A. In a survey, LBD caregivers expressed concerns about fear of future (77 percent), feeling stressed (54 percent), loss of social life (52 percent), and uncertainty about what to do next (50 percent).[6] Caregivers reported moderate to severe burden; 80 percent felt the people around them did not understand their burden, and 54 percent reported feelings of isolation, with spousal caregivers reporting more burden than nonspousal caregivers.[6,7] Two-thirds of the caregivers reported medical crises requiring emergency services, psychiatric care, or law enforcement.[6]

Studies of LBD caregivers have demonstrated higher subjective burden compared to that found in studies of individuals caring for those with Alzheimer's.[7] The caregiving burden affects physical health, emotional health, and relationships with friends and family. A unique feature of the caregiving burden in dementia with Lewy bodies has been that caregivers question their performance in providing care. It has been suggested that the lack of awareness about dementia with Lewy bodies compared to Alzheimer's and other forms of dementia limits the access of caregivers to information and support, leading to emotional isolation and increasing perception of burden.[7]

Even when the diagnosis is made, there are no specific medications approved for the treatment of LBD. Instead, medications for Alzheimer's and Parkinson's are often used to treat individual symptoms. LBD's psychotic symptoms, such as visual hallucinations and behavioral disturbances in dementia, can often be distressing. Antipsychotic medications, often used in other dementias for psychotic symptoms and behavioral disturbances, require special caution in LBD. LBD patients may have an extreme sensitivity to antipsychotic medications, even at low doses, resulting in sudden confusion, worsening of rigidity and immobility, and, in rare cases, a lethal syndrome termed "neuroleptic malignant syndrome," which is accompanied by high fever.[23] LBD patients may also display paradoxical reactions to medications, whereby they become activated and awakened by commonly used sleep aids.[24]

If medications are employed for LBD patients, a complex and delicate degree of balance is required. Drugs that are commonly used to treat the rigidity and immobility of Parkinson's cannot be used as liberally in dementia with Lewy bodies, since they tend to worsen hallucinations. Similarly, drugs used in treating urinary symptoms tend to cause confusion and worsen memory and attention.

Steps Ahead

Lewy bodies were first described in the early 1900s by Friederich H. Lewy while researching Parkinson's disease. However, the first case of LBD was not described until 1961, with the first set of clinical criteria put forth in 1996. One reason LBD research has lagged behind that focusing on Alzheimer's and Parkinson's for decades is due to an earlier notion that it was a rare disease. It wasn't until the development of a staining technique in the late 1990s that researchers learned how much more common LBD is than previously thought.

There is currently a dearth of drug development in LBD, which is urgently needed because of the limited availability of medications that do not exacerbate symptoms. With fewer unique disease targets identified to date, it is increasingly difficult to secure federal research funding in the current, very competitive climate. Another important problem that delays clinical trials for new medications is the current difficulty in diagnosing LBD. At the present time, it would be difficult to register a clinical trial with the FDA because it is unclear that clinicians could correctly make a diagnosis—this would affect both clinical trial recruitment as well as identifying appropriate patients for therapy.

There is good news on the horizon, though, in that there are some compounds being developed for cognition/dementia in Parkinson's, which is easier to diagnose due to the earlier broad Parkinson's diagnosis. Drugs that become FDA-approved for PDD may be ideal candidates for clinical trials in LBD down the road. New Alzheimer's drugs may also hold potential for LBD, due to the strong overlap of Alzheimer's pathology in LBD.

Like Alzheimer's and Parkinson's, LBD is a complicated illness and its mysteries won't be easy to solve. Because all three illnesses have similar symptoms, LBD is frequently misdiagnosed and underdiagnosed. This delays treatment with currently available medications, places individuals with LBD at risk for exposure to severe medication side effects, and delays testing of promising therapeutics.

One way to improve diagnoses is to use standardized scales that increase the accuracy of detecting LBD, such as the Lewy Body Composite Risk Score,[12] containing 10 items that may accurately and reliably identify LBD as the cause of cognitive impairment. A number of research laboratories are now able to measure alpha-synuclein in cerebrospinal fluid; a clinically available test may be just on the horizon. A number of labs are actively trying to develop imaging markers for alpha-synuclein, similar to what has been done for amyloid imaging in Alzheimer's, but these are probably three to five years away.

These advances in diagnostics, however, hold great promise. For now, it is imperative to establish accurate diagnosis and institute prompt treatment. We also need to raise awareness of LBD for the public and for health-care providers through continuing medical education, increase advocacy for patients and caregivers by supporting groups such as the Lewy Body Dementia Association, and increase research funding to understand the causes and develop the novel therapies needed to address this under-recognized but all too common cause of dementia.

11

Getting High on the Endocannabinoid System

By Bradley E. Alger, Ph.D.

Bradley Alger, Ph.D., is a professor in the Department of Physiology at the University of Maryland School of Medicine. He received his Ph.D. in experimental psychology from Harvard University in 1977. In 1981, he was appointed assistant professor at Maryland, and was promoted to professor in 1991, with a secondary appointment in psychiatry awarded in 2004. His long-term research interests are on the regulation of synaptic inhibition in the brain. In the early 1990s, Alger and Thomas Pitler co-discovered (in parallel with the laboratory of A. Marty) "depolarization-induced suppression of inhibition (DSI)", the first thoroughly characterized and widely accepted instance of retrograde signaling in the brain. DSI was eventually found to be mediated by endocannabinoids by the laboratories of R. Nicoll and M. Kano, and is the first instance of a physiological process carried out by endocannabinoids. In all, Alger's group has published more than 100 research papers on the regulation of inhibition, focusing mainly on DSI and endocannabinoids in the past two decades.

The endogenous cannabinoid system—named after the plant that led to its discovery—is one of the most important physiologic systems involved in establishing and maintaining human health. Endocannabinoids and their receptors are found throughout the body: in the brain, organs, connective tissues, glands, and immune cells. With its complex actions in our immune system, nervous system, and virtually all of the body's organs, the endocannabinoids are literally a bridge between body and mind. By understanding this system, we begin to see a mechanism that could connect brain activity and states of physical health and disease.

CANNABIS, DERIVED FROM A PLANT and one of the oldest known drugs, has remained a source of controversy throughout its history. From debates on its medicinal value and legalization to concerns about dependency and schizophrenia, cannabis (marijuana, pot, hashish, bhang, etc.) is a hot button for politicians and pundits alike. Fundamental to understanding these discussions is how cannabis affects the mind and body, as well as the body's cells and systems. How can something that stimulates appetite also be great for relieving pain, nausea, seizures, and anxiety? Whether its leaves and buds are smoked, baked into pastries, processed into pills, or steeped as tea and sipped, cannabis affects us in ways that are sometimes hard to define. Not only are its many facets an intrinsically fascinating topic, but because they touch on so many parts of the brain and the body, their medical, ethical, and legal ramifications are vast.

The intercellular signaling molecules, their receptors, and synthetic and degradative enzymes from which cannabis gets its powers had been in place for millions of years by the time humans began burning the plants and inhaling the smoke. Despite records going back 4,700 years that document medicinal uses of cannabis, no one knew how it worked until 1964. That was when Yechiel Gaoni and Raphael Mechoulam[1] reported that the main active component of cannabis is tetrahydrocannabinol (THC). THC, referred to as a "cannabinoid" (like the dozens of other unique constituents of cannabis), acts on the brain by muscling in on the intrinsic neuronal signaling system, mimicking a key natural player, and basically hijacking it for reasons best known to the plants. Since the time when exogenous cannabinoids revealed their existence, the entire natural complex came to be called the "*endo*genous cannabinoid system," or "endocannabinoid system" (ECS).

THC is a lipid, but in 1964, known or suspected neurotransmitters and neuromodulators were water-soluble molecules—peptides, amino acids, or amines—not lipids. Ordinary neuroactive agents interact with cells by binding to specific proteinaceous receptor molecules that are part of the cell surface. Each receptor has an intricate structural pocket into which a particular neurotransmitter fits. The interaction triggers the biochemical and biophysical reactions that affect the physiological properties of the cell. Lipids avoid water, and individual lipid molecules might simply drift freely around in a compatible lipophilic environment, such as the cell surface membrane, without having much to do with proteins. How could they influence neuronal behavior?

The best scientific guess at the time was that molecules such as THC would owe their psychotropic actions to "membrane fluidizing" properties, a vague notion that would not explain specificity of action, among other things. Nevertheless, strong evidence that THC and similar synthetic molecules could bind tightly to specific sites in the brain emerged,[2] implying that THC does indeed work through true receptors. This hypothesis was confirmed in 1990 with the isolation and cloning of the first cannabinoid receptor, CB1,[3] and later of CB2.[4]

In the central nervous system (CNS), CB1 is by far the predominant form, although it also exists outside the CNS; CB2 is primarily found outside the CNS, and is associated with the immune system. Both receptor subtypes are 7-transmembrane domain macromolecules of the "G-protein-coupled" class. Unexpectedly, CB1 turned out to be one of the most abundant G-protein-coupled receptors in the brain. It was immediately obvious that CB1 and CB2 must partner with an endogenous ligand, a natural agent for which they would normally act as the proper receptors. They did not evolve to react with rarely ingested, plant-derived chemicals. Indeed, Mechoulam's group isolated an arachidonic acid derivative

(N-arachidonoylethanolamide, "anandamide") that activated CB1,[5] and a second endogenous CB1 ligand two-arachidonolyl glycerol (2-AG) was later discovered.[6,7]

These endocannabinoids are the major physiological activators of CB1 and CB2, yet they are not standard neurotransmitters. For one thing, like THC, they are lipids, and brain cells, mainly neurons, are surrounded by an aqueous solution, an inhospitable environment for an intercellular lipid messenger. More surprisingly, endocannabinoids go against the flow of typical chemical synaptic signaling. A neuron that releases a chemical neurotransmitter (say, GABA or glutamate) is designated as "pre-synaptic"; the target neuron that expresses receptors for that neurotransmitter is "post-synaptic." Endocannabinoids, however, are synthesized and released from post-synaptic cells, and travel backward (in the "retrograde" direction) across the synapse, where they encounter CB1s located on adjacent nerve terminals.[8,9] Physiologically, CB1Rs act as communications traffic cops. Precisely positioned in synaptic regions,[10] they inhibit the release of many excitatory and inhibitory neurotransmitters. Thus, by releasing endocannabinoids, postsynaptic target cells can influence their own incoming synaptic signals.

CB1 is densely located in the neocortex, hippocampus, basal ganglia, amygdala, striatum, cerebellum, and hypothalamus. These major brain regions mediate a wide variety of high-order behavioral functions, including learning and memory, executive function decision making, sensory and motor responsiveness, and emotional reactions, as well as feeding and other homeostatic processes. Within neuronal circuits, suppression of excitatory transmitter release tends to dampen excitation, while suppression of inhibitory transmitter release favors neuronal network excitation. Given the enormous complexity of the brain, the endocannabinoid system could affect behavior in an almost limitless number of ways: Simple generalizations of what will happen when CB1 receptors are globally turned on or off are not feasible. The challenge for developers of cannabinoid-based medicines is to find beneficial ways to exploit this powerful yet convoluted feedback system.

From a therapeutic point of view, the near ubiquity of the endocannabinoid system has good news/bad news implications. Good news because it offers explanatory power—the ability to make sense of numerous yet quite different aspects of neural processing involving the endocannabinoid system in normal brains and, conversely, to offer insight into a variety of maladies that accompany its dysfunction. Bad news because wide heterogeneous dispersion greatly complicates the task of targeting this system for specific therapeutic purposes. Side effects are therefore common and problematic.

CB1 and Obesity

Obesity is a serious worldwide health concern. An attempt to develop an endocannabinoid system–based strategy to solve it provides a textbook example of the promise and the problems involved. The feeding control centers in the hypothalamus express high concentrations of CB1. These receptors are responsible for "the munchies," the craving for food that is stimulated by cannabis use. But they also prompt the normal desire to eat. Preventing the activation of hypothalamic CB1s should decrease eating. In addition, CB1 receptors outside the brain regulate energy metabolism in the liver and fat tissue,[11] and pharmacologically blocking these peripheral receptors in animal studies results in less body weight gain even when the same amount of food was eaten.[12] Researchers at the pharmaceutical company Sanofi-Aventis gave the CB1 antagonist rimonabant to obese individuals in multi-year, multi-thousand-patient trials and obtained stunning results. The drug worked brilliantly; patients lost weight and girth. Negative side effects (depression, anxiety, and nausea) occurred in 10 percent of the users, but they were not life-threatening and the risks were deemed worth the rewards. Rimonabant (marketed as "Acomplia®"among other names) became readily available in 56 countries in 2006, and Sanofi's stock soared. When approached to approve sale in the United States, however, the Food and Drug Administration (FDA) was skeptical and asked for more information about the drug's performance after the clinical trials had ended.

The trials had excluded people who were susceptible to psychiatric illness, including depression. What was the experience like in the real world, where many obese patients also suffer from mental disturbances? The answer was alarming: The incidence of serious depression, including suicidal ideation, bouts of nausea, stress, and anxiety was markedly higher than in the trials. The beneficial effects of rimonabant and its downsides both arose from the same source. Blocking CB1 in the hypothalamus was beneficial because it diminished the desire to eat, but the drug, which was given orally, blocked CB1 throughout the body, including in those brain regions where the endocannabinoid system regulates emotion and vomiting reflexes, among others. Which effects predominated was a matter of individual variation, and it had to be assumed that widespread use of rimonabant would put many people at risk for serious adverse consequences. The FDA disapproved its distribution in the United States, and as reports of bad outcomes increased among patients in other countries, it was soon withdrawn from the market. As a result, Sanofi's stock came back to earth.

Inhibit vs. Stimulate

Some conditions, such as chronic pain, spasticity, anxiety, and the wasting syndrome associated with chemotherapy and AIDS, can be alleviated by cannabinoids, and therefore therapeutic approaches would involve activating, not inhibiting, CB1. For example, people self-medicate with cannabis to relieve anxiety. The endocannabinoid system helps us deal with traumatic life experiences as a part of a normal coping mechanism—to forget it and leave the past behind. Neuroscientists use animal models, often the "fear conditioning" test, to investigate the development of anxiety. This is a Pavlovian training procedure in which a mildly unpleasant stimulus (a brief electric shock to the wire floor grid on which a rat or mouse is standing) is paired with a neutral tone, audible though not loud. The shock causes the animals to freeze in position—the typical response of small rodents to threatening stimuli. When the tone is sounded alone, it elicits a bit of curiosity, then soon is ignored. When the tone repeatedly precedes and accompanies the foot shock, the animal comes to recognize it as a bad omen and eventually responds when the tone first sounds even when the shock no longer occurs. The animal has acquired conditioned fear.

Normal coping includes dissipation of the bad memories evoked by the tone (actually learning that the tone is no longer threatening), a process called "extinction," which enables animals to cease paying the high costs of pointless responding. Mice genetically engineered so that they do not have CB1 receptors readily acquire the fearful response, but cannot forget it as easily as do normal mice.[13] These mutant mice continue to respond fearfully to the tone alone, even though it no longer signals that the shock is coming, suggesting that activation of the endocannabinoid system is an essential component of the coping mechanism. Failure to extinguish learned fearful responses may underlie posttraumatic stress syndrome (PTSD) in humans. Stimulation of the endocannabinoid system could be useful in the treatment of PTSD, as it is for treatments of cachexia and spasticity.

Inhalation vs. Digestion

The most direct route of THC administration is by smoking marijuana or other forms of cannabis. Yet purified, FDA-approved medicinal preparations of THC are available in pill form (dronabinol, pure THC marketed as Marinol®, and the analog nabilone, sold as Cesamet® in Canada). If THC is the active agent in cannabis, and approved, orally effective THC medications exist, why the impetus for medical marijuana? In addition to avoiding all of the legal, political, and social hassles (pot purveyors occasionally being unsavory characters), avoiding inhalation of particulates in smoke is highly desirable on its own. Why not just take a pill?

There are several reasons that some patients prefer puffing over swallowing. One quantitatively minor factor is potential lethality. It is possible to get a fatal overdose by swallowing too many THC pills at once, whereas documented evidence of death simply from smoking too much cannabis does not seem to exist.

More common factors are speed and predictability of action, and degree of patient control. Pills must enter the digestive system, where the rate of entry of THC into the bloodstream is slow and dependent on the state of gastric filling. It can take more than an hour for the full influence of ingested THC to be exerted on the brain, and even that time will vary depending on the timing and contents of one's last meal. In contrast, it takes only 20 to 30 seconds for inhaled THC to reach the brain from the lungs, and its peak effects are achieved within a few minutes. For someone suffering nausea (itself a significant impediment to the swallowing of medicine or anything else) or chronic pain, the choice is often not a difficult one.

The third factor, controllability, is another serious concern. Once a pill is swallowed, the full dose is on its way with its time-course and side effects to be played out inexorably, governed by the rates of absorption and clearance of the drug from the body. An effective dose that has tolerable side effects in a robust middle-aged man may be too much and have intolerable psychotropic side effects in a slight, elderly woman seeking appetite stimulation to counter the weight loss associated with cancer chemotherapy. With inhalation, patients become adept at sensing and adjusting their intake of THC via smoking (just as people become good at titrating their blood levels of nicotine when smoking tobacco). Because smoked THC enters the brain so quickly, patients can readily detect its presence and adjust their dosing to the level that they need by inhaling less or more. A significant downside to inhalation is that the by-products of burning plant material, particulate and chemical, are taken in and can irritate the mucous membranes of the mouth and lungs. Even though most marijuana smokers do not smoke as much as a pack-a-day tobacco smoker does, bronchitis and the buildup of carcinogenic tars in the lungs do occur in heavy users. Studies of the occurrence of chronic obstructive pulmonary disease (COPD) from cannabis smoking are inconsistent, though. Finally, while generally anxiety-relieving (anxiolytic) in low doses, THC can provoke anxiety and paranoia in high doses, responses that seem exacerbated with inhalation, probably because it acts so quickly.

Some of the drawbacks of smoking cannabis may be circumvented by the use of vaporizers somewhat similar to "e-cigarettes" (electronic cigarettes) that use heating elements to vaporize a liquid nicotine solution. Cannabis vaporizers heat the plant material so that volatile compounds, such as THC, are given off before actual burning and the associated release of particulates, toxins, and carcinogens occurs. Such devices deliver about as much THC as is found in smoke and are often better tolerated than smoking, although irritation of the mouth and throat are occasional problems. Like e-cigarettes, the designs, efficacy, safety, regulation, and legality of these devices are in flux, but they do provide a potential option for cannabis users who prefer inhalation.

Variation (polymorphisms) among people in the genes encoding CB1 receptors and other endocannabinoid system components affect their cannabinoid drug sensitivity,[14,15] as well as their susceptibility to disorders related to disturbances of the endocannabinoid system. Links between CB1 polymorphisms and schizophrenia, autism-spectrum disorders, and PTSD have been suggested but remain controversial. Sorting these relationships out is an important task, since the information gained will contribute to the future ideal of personalized medicine.

Would Having an Entourage Help?

A final reason for the popularity of smoking over the purified oral THC preparations is subtle and not well understood. For many people, pure THC in pill form is aversive; the unpleasant sensations, "dysphoria," cause patients to not take their pills. Smoking cannabis is less offensive for some of these patients, suggesting that something besides THC is involved. THC is the only psychotropic cannabinoid, but one or more of the nonpsychotropic cannabinoids could modulate or soften the impact of pure THC in several ways: They might act as part of an "entourage,"[16] unable to activate CB1 themselves but capable of modifying THC's ability to do so. Alternatively, nonpsychotropic cannabinoids might influence other components of the endocannabinoid system (synthesis, uptake, or degradation), and thus alter availability of endocannabinoids, which compete with THC for access to CB1, and thereby indirectly tweak THC's actions.[17] But interactions with the endocannabinoid system are not the only possibilities. Non-psychotropic cannabinoids can affect conventional neurotransmitter receptors and ion channels that are entirely unrelated to the endocannabinoid system.[18] (They are "cannabinoids" because they come from the cannabis plant, not because they necessarily have anything to do with CB1, CB2, or the ECS in general.)

Cannabidiol (CBD) is a major nonpsychotropic cannabinoid, and is almost as abundant as THC. Interestingly, while the CBD:THC ratio varies in different strains of cannabis, the total amount of cannabidiol plus THC across strains is roughly constant. The more THC, the less cannabidiol, and vice versa. The proportion of CBD:THC is selected for in cannabis plant-breeding programs. Cannabidiol can inhibit CB1 (and CB2) directly, and this may diminish THC's CB1-mediated undesirable actions,[17] which are dose-related. For example, cannabidiol blunts the anxiogenic and psychotropic side effects of THC. In addition to synergistic actions, cannabidiol by itself is anxiolytic,[18] and can reduce inflammation and blood pressure.[19] A mucosal spray, Sativex® (GW Pharmaceuticals), a botanical extract of cannabis plants, has a standard CBD:THC ratio of 1. In Canada, the United Kingdom, and other countries (not yet the United States), Sativex® is available for the treatment of the pain and spasticity of multiple sclerosis.

The anticonvulsant properties of cannabis have been known for centuries. A dramatic account of such action recently received widespread media coverage.[20] A young child suffering from an intractable form of childhood epilepsy called Dravet syndrome had been unsuccessfully treated with a battery of epilepsy therapies for years since her first seizure at three months old. By age 5, she was having up to 300 seizures per day, and experiencing mental and physical developmental stagnation. Her prospects were grim and her parents desperate. With the approval of two doctors, they tried adding an oil extract of cannabis to her food. Amazingly, her seizures immediately dropped to a few per month, an improvement that has persisted for a year, and her normal development resumed.

A notable feature of this case, which has been repeated in other similarly afflicted children, is that her cannabis extract is from a strain (called "Charlotte's Web") that is very low in THC and high in cannabidiol. To what extent this positive outcome is attributable to the low THC, the high cannabidiol, or the combination of the two is unknown. A different non-psychotropic cannabinoid, cannabidivarin, reduces seizures independently of CB1 in animal models, and this property is not improved by the presence of THC.[21]

Turning On (or Off)

CB1 receptors exist on nerve fibers outside of the central nervous system, and there they also direct communications traffic. Psychotropic side effects of cannabis are caused exclusively by turning on or off brain CB1s. Therefore one strategy is to develop CB1 agonists or antagonists that can be given orally but that do not cross the blood-brain barrier (a membranous cellular fence that bars certain chemicals present in the circulation from getting into the brain). CB1s in fat and other tissues are thought to contribute to obesity, and a peripherally restricted CB1 antagonist could be beneficial in weight control. Conversely, cannabinoids are good pain relievers that work in part by stimulating CB1s on peripheral pain sensory neurons. When activated, these CB1s block transmission of the pain signals to the brain—basically what topical anesthetics like novocaine do—and pain signals unable to reach the brain are not felt. CB1 agonists or antagonists that are restricted from the brain could be quite useful in conditions that do not arise from within the central nervous system.

What about manipulating other components of the endocannabinoid system? Rather than stimulating CB1 with drugs, the endocannabinoids can be pressed into service artificially. Once released, endocannabinoids, like other chemical messengers, are quickly taken back up into cells or otherwise inactivated, which preserves the integrity of the signaling process. Inhibiting uptake and degradation therapeutically offers the advantage of increasing the endocannabinoid levels, and thereby activating CB1, in those regions in which the messengers are already being mobilized by brain activity itself. Rather than indiscriminate activation of CB1s everywhere for long periods of time, only certain groups of receptors would be activated and only when and where called for naturally. With a drug that inhibits the enzyme (fatty-acid amide hydrolase, FAAH) that inactivates the endocannabinoid, anandamide (but not 2-AG), levels increase,[22] and an analogous approach inhibits the major degradation enzyme for 2-AG, monoglyceride lipase (MGL) and 2-AG levels rise.[23] Elevations in endocannabinoids in this way can have beneficial effects.[24] Unfortunately, there are still problems: In addition to activating CB1, anandamide turns out to be an excellent activator of a another receptor, TRPV1,[25] a noncannabinoid receptor that actually heightens anxiety, so globally elevating anandamide has complex effects.[18] Drugs that inhibit both FAAH and TRPV1 could be helpful in some cases.[26] Meanwhile, globally elevating 2-AG by decreasing its breakdown overloads the endocannabinoid system, which responds by causing a protective shutdown, or down regulation, of many CB1s in the brain.[27] This is counterproductive if the goal is stimulation of the endocannabinoid system.

An encouraging development along these lines is that the peripheral pain signals can be quashed by raising anandamide and 2-AG levels only near the site of origin (a rat's paw), where a painful stimulus was given.[28] This means that the local peripheral CB1 and CB2 receptors in the paw were effectively turned on by the elevation in endocannabinoid levels resulting from prevention of their breakdown. In this case, pain relief free of psychotropic side effects should be possible with degradative enzyme blockers designed to stay out of the central nervous system.

Finally, a possibility that has gotten little attention is the targeting of conventional neurotransmitter systems that stimulate the production of endocannabinoids. For example, glutamate is the major excitatory neurotransmitter in the brain, and one subtype of glutamate receptors (group I mGluRs) potently mobilizes endocannabinoids.[29,30] A genetic disease that causes mental retardation, fragile X syndrome, has long been associated with excessive activity at the same glutamate receptors,[31] which could be related to the excess production of endocannabinoids at inhibitory synapses in a mouse model of the disease.[32] Perhaps combining modest inhibition of both CD1 and group I mGluRs would be a way of tapping the therapeutic potential of the ECS, while avoiding some of its problems.

What Is in Store?

The endocannabinoid system is powerful and nearly ubiquitous in the nervous system. The cannabinoid receptors dispersed throughout many brain regions are responsible for regulation of numerous aspects of neuronal activity, and account for the bewildering variety of behavioral and psychological effects caused by THC. Depending on the nervous system regions and maladies involved, either stimulating or inhibiting the endocannabinoid system could have beneficial effects. A great deal of attention is being given to incorporating nonpsychotropic cannabinoids into medicinal preparations, although in most cases the actual effects of these agents on the nervous system are unknown. For some purposes, drugs that are restricted to acting on peripheral cannabinoid receptors, and are prevented from entering the central nervous system, could be effective. Finally, therapeutic strategies aimed at developing regionally selective targeting of endocannabinoid system components, perhaps in combination with agents that affect conventional neurotransmitter systems, or nonpsychotropic cannabinoids, offer promise for future advances.

12

Migraine and Sleep
New Connections

By Andrew H. Ahn, M.D., Ph.D., and Peter J. Goadsby, M.D., Ph.D.

Andrew H. Ahn, M.D., Ph.D. is a neurologist with a clinical subspecialty in the evaluation and management of headache and facial pain, and associate editor of the journal *Headache: The Journal of Head and Face Pain.* His research has received support from the Howard Hughes Medical Institute Postdoctoral Fellowship for Physicians, the Mentored Career Development Award from the National Institutes for Health, and awards from the American Academy of Neurology, the American Headache Society, and the International Headache Society. Ahn received both his M.D. and Ph.D. from Harvard University and trained as a postdoctoral fellow and resident at the University of California-San Francisco.

Peter Goadsby, M.D., Ph.D. is a neurologist who specializes in the diagnosis and treatment of headache disorders, including migraines and cluster headache and other forms of chronic daily headache. He is director of the University of California-San Francisco Headache Center and a leading headache expert and researcher. Goadsby came to UCSF from the Institute of Neurology at University College London, National Hospital for Neurology and Neurosurgery and Hospital for Sick Children. He earned a medical degree at the University of New South Wales-School of Medicine, completed a residency in neurology at the Prince of Wales Hospital, and completed a fellowship in neurology at the National Hospital for Nervous and Mental Diseases in London. He completed postdoctoral work in laboratories in New York and Paris.

This story focuses on an important link between migraine and sleep patterns. A better understanding of the relationship among the body's circadian rhythms, the brain's hypothalamus, and a mutated gene holds enormous promise of improved care for the more than 36 million Americans who experience migraine (three times more common in women) and the countless number of people suffering from familial advanced sleep phase syndrome (FASP).

BOTH AMERICAN AND BRITISH CULTURES extol the moral virtues of the early riser. In 1670, the English naturalist John Ray noted in his *A Complete Collection of English Proverbs* that:"The early bird gets the worm." Next came the 1735 edition of Benjamin Franklin's *Poor Richard's Almanack*, which also clearly indicated that such virtue has its material and economic rewards:"Early to bed and early to rise makes a man healthy wealthy, and wise."

From Thomas Jefferson's "The sun has not caught me in bed for 50 years" to Henry David Thoreau's "To him whose elastic and vigorous thought keeps pace with the sun, the day is perpetual morning," such are the virtues attributed to the "morning lark," arising early and singing productively at the day's earliest hours. Even the songbird to which we refer is intimately associated with the earliest morning hours, such as in Chaucer's *The Knight's Tale*, "the bisy larke, mesager of day." And so much the worse for us poor "night owls" who find our most productive hours only after everyone else has gone to bed.

But before one fully endorses the virtues of "morning lark" versus "night owl," consider that biologists have found that either trait may be less a matter of moral virtue and more a matter of brain physiology. Scientists interested in circadian rhythms—the body's own daily clock—have noted durable differences in activity pattern that distinguish the two. The circadian rhythm is a biological clock that sets itself to daylight conditions, and regulates many physiological signals, such as a morning surge of activity-promoting corticosteroids from the adrenal glands and a nighttime peak of the sleep-promoting melatonin from the pineal gland. The clock implicates the hypothalamus, the phylogenetic, ancient part of the brain responsible for keeping the body working steadily, on an even keel. Its so-called *homeostatic* functions cover a range of critical physiological functions, such as body temperature, blood pressure, feeding and satiety, blood glucose, and the regulation of sex hormones.

Our Internal Clock

Circadian biologists had clearly established that even in the absence of daylight cues, there persists an internal clock. And while humans have evolved an internal circadian rhythm that corresponds closely to the 24 hours of the day cycle, there is a range of normal circadian lengths, which include either somewhat shorter or longer rhythms. Those with shorter rhythms have circadian rhythms that arrive more quickly to the end of their day, and so it is that those people with extremely short circadian cycles are said to have so-called advanced sleep phase (ASP). Those with ASP have an internal rhythm that drives them to bed early, and are likely to rise bright and early after a full night's sleep, ready to take on the new day.

On the other hand, those with a long circadian rhythm, referred to as delayed sleep phase, are night owls. Their drive to sleep turns on later in the night, and they are likely to have trouble getting up in the morning because, having gone to bed late, they have not had a full night's rest when most people are starting their day. The circadian length also tends to change over a lifespan, being somewhat longer as an adolescent and early adult, and growing shorter with age.

Over the last decade, the recognition that the *extreme* morning lark trait can run in families presented geneticists Louis Ptácek and Ying-Hui Fu with the opportunity to dissect the molecular pathways related to circadian rhythm in humans. This led to the revelation that a mutation of the hPer2 gene—the human homologue of a family of *period* genes in fruit flies—is a key player in how the hypothalamus keeps track of this daily rhythm that drives our innate pattern of wakefulness and activity. The fundamental significance of a common biochemical pathway underlying circadian rhythm throughout evolution cannot be overstated.

From a more personal perspective, it is also striking how the lives of those with a familial form of severe ASP (FASP) are deeply etched by the burden of their hereditary condition. With a drastically short cycle, their physiologic "day" runs out early, and these people (and their affected family members) are often driven to sleep by 7:30 pm. Those affected report a lifelong pattern of activity that involves absolutely no night-life, often springing to consciousness after a very full night's rest at 4:30 am. It was further good fortune to find that two other family members with FASP had alterations of the gene CK1delta, an enzyme that is also implicated in the activity of the hPer2 gene in the hypothalamus.

From Lark to Spark

Enter migraine, a common, complex brain disorder whose biology needs to be better understood. Migraine is a common type of headache that may occur with symptoms such as nausea, vomiting, or sensitivity to light. In many people, a throbbing pain is felt only on one side of the head. Some people who get migraines have warning symptoms called an aura, before the actual headache begins.

The connection made between migraine and circadian rhythm is a brilliant example of how different fields with no apparent connection can receive a completely unscripted and unforeseen spark of insight, waiting only for the prepared mind to grasp the opportunity. That inspired spark took place when migraine researcher Robert Shapiro realized that his patient, who was seeing him for migraine with aura, was a member of an extended Vermont family of extreme morning larks.

A study of the 14 members of the family showed that the CK1delta gene implicated in FASP was also connected to migraine with aura. Five who had identical mutations in the CK1delta gene also met the diagnostic criteria for migraine. The gene was then sequenced in blood samples from 70 additional families with the rare sleep disorder. One family had a slightly different mutation in the CK1delta gene. In this family, too, all five members with CK1delta gene mutations had migraine, aura without migraine, or probable migraine. The researchers later showed that the CK1delta mutations in both families reduced the enzyme's activity.

While it was now clear that CK1delta had a causal link to FASP through its regulation of the hPer2 protein, whether the relationship to migraine was merely coincidental or truly linked by a causal mechanism required further arduous work, drawing inferences of what can be known about migraine from animal models. These models were at best an approximation of a complex neurological disorder that includes pain and a host of sensory, motor, autonomic, gastrointestinal, and affective symptoms.

The details of this association (from a cellular or neurotransmitter point of view) still require further work, but one reason for all the buzz among neuroscientists is that this connection advances the likely role of the hypothalamus in migraine. In addition to the multitude of clinical associations between migraine and sleep, many other clinical features of migraine call attention to the connection between migraine and the hypothalamus.

In the lives of those affected by migraine—so called migraineurs—sleep is a key factor. In fact, any disturbance from the normal routine of sleep is often identified as a trigger of migraine, such as staying up too late, getting up earlier than usual, a change in activity pattern due to shift work, jet-lag from crossing many time zones, or even oversleeping. Accordingly, any persistent disturbances of regular sleep can be a cause of frequent or difficult-to-control migraine, such as an erratic sleep schedule, frequent changes in activity due to shift work, frequent interruptions of sleep through the night, a chronic illness that can disturb sleep quality, or the presence of any one of a number of sleep disorders, such as obstructive sleep apnea.

As a result, an important component of effective management for those with frequent or severe migraine is to identify the trigger factors and aggressively neutralize them. The major challenge is convincing the migraineur to make the often disruptive life adjustments that are needed to manage these triggers, and to give the changes enough time to assess their efficacy.

Migraine Trigger Factors

There are a range of well-established trigger factors for migraine that reference the hypothalamus. These are present in many but not all migraineurs, and even for those who identify triggers, they do not represent absolute triggers (every time a migraine) but are merely associations. More often than not, these factors can be thought of as cumulative physiological challenges that require the attention of the hypothalamus to normalize the physiological stresses that they represent. These can include an acute psychological stress, a skipped meal, overexertion with overheating, or lack of sleep.

This background connects with further recent developments implicating the role of the hypothalamus in the regulation of pain. Among many such examples, some of which may have implications for therapy, one recent line of evidence has shown that an antagonist to orexin, a peptide hormone associated with the regulation of feeding behaviors, can also have properties that suppress physiological responses to sensory stimuli in the cranial dura mater.

Also, oxytocin, a neuropeptide sometimes associated with social and nurturing behaviors in mammals (the so-called love hormone), is thought to act on another hypothalamic regulator of blood pressure, the vasopressin receptor, which geneticist Jeffrey Mogil and colleagues linked to a genetic variant in the human population that can change responses to pain in men presented with a high-stress situation. Again, interestingly, preliminary evidence suggests that commonly available veterinary oxytocin can be administered through the nose in humans to cross the blood brain barrier, and preliminary evidence suggests that it may also be an effective target for headache. Other neurotransmitters of the hypothalamus are also of great interest to migraine pathophysiology. These include melatonin and the so-called pituitary adenylate cyclase activating peptide (PACAP), both of which are related to hypothalamic function.

Another independent line of neuroimaging data suggests that the hypothalamus may have an early causative role in the pathophysiology of migraine. In this study nitroglycerin was used as an experimental stimulus to trigger what would otherwise be considered typical spontaneous migraine attacks in people who also normally have early warning symptoms prior to their attacks. The activation of blood flow to the posterolateral hypothalamus during the warning phase, compared to baseline figures, implicates the early role of the hypothalamus in migraine attacks. Future study may challenge the homeostatic functions of the hypothalamus.

BOOK REVIEWS

13

Ain't No Mountain High Enough

How Children Succeed: Grit, Curiosity, and the Hidden Power of Character

By Paul Tough
Reviewed by Silvia A. Bunge, Ph.D.

Silvia A. Bunge, Ph.D. is an associate professor and vice chair of the Department of Psychology and an associate professor in the Helen Wills Neuroscience Institute at the University of California at Berkeley. Her other affiliations at UC Berkeley include the Institute of Human Development and the Research in Cognition and Mathematics Education program. Dr. Bunge directs the Building Blocks of Cognition Laboratory, which draws from the fields of cognitive neuroscience, developmental psychology, and education research, and seeks to better understand both negative and positive environmental influences on brain and cognitive development. Through her research and public service, she hopes to promote academic readiness among children at risk for school failure.

There is no obstacle in the path of young people who are poor or members of minority groups that hard work and preparation cannot cure.

—Barbara Jordan, a leader of the civil rights movement

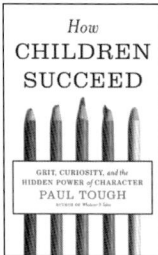

PAUL TOUGH'S WORK REFLECTS his enduring concern about the achievement gap in our society between children from the poorest and wealthiest families in the United States. In his first book, *Whatever It Takes: Geoffrey Canada's Quest to Change Harlem and America*, he chronicled the creation of the Harlem Children's Zone, Inc., an organization that has received national acclaim for its efforts to pull children out of the intergenerational cycle of poverty.

Tough seeks to dispel the notion that children from underserved communities are doomed to fail. The basic premise is that, regardless of a child's IQ, she or he can excel through hard work and perseverance—if given proper encouragement and opportunity. Rather than dwelling on sobering national statistics, which do little to move even the most well-intentioned reader to action, he focuses on the human element. He does this by showing what is possible through the success stories of individual children, teachers, and schools and also by introducing us to top scientists who explain first-hand the significance of their and others' research on child development and education.

Each of the three blurbs on the jacket of his latest effort, *How Children Succeed*, praises a different aspect of the book. Annie Murphy Paul writes in *The New York Times Book Review* that it "illuminates the extremes of American childhood: for rich kids, a safety net drawn so tight it's a harness; for poor kids, almost nothing to break their fall." Certainly, this harsh disparity serves as the impetus for the book, but it is not front and center. A reader hoping for an in-depth analysis of inequalities in the American educational system and promising approaches toward school reform would likely be better off picking up a copy of Linda Darling-Hammond's book, *The Flat World and*

Education: How America's Commitment to Equity Will Determine Our Future.

Charles Duhigg, author of *The Power of Habit,* commends Tough for introducing "new, powerful ideas about how to help children thrive, innovations that have transformed schools, homes, and lives." Indeed, Tough focuses on some of the most promising ideas and practices of the past decade. He reviews the work of researchers studying key factors that influence academic achievement, including adverse childhood experiences (e.g., Nadine Turke-Brown, Michael Meaney), self-control (e.g., Walter Mischel, Clancy Blair), and motivation and mindset (e.g., Angela Duckworth, Carol Dweck). He also takes us into the schools, showcasing the legendary KIPP Academy in the South Bronx and delving into the life and work of a highly effective chess teacher in Brooklyn named Elizabeth Spiegel.

Finally, Alex Kotlowitz, author of *There Are No Children Here,* notes that at the book's "core is a notion that is electrifying in its originality and its optimism: that character—not cognition—is central to success, and that character can be taught." Indeed, this is the central premise of Tough's book, and the one that best differentiates it from others. While I share Tough's optimism regarding the potential to alter the life trajectories of underserved children, I believe that this central message is flawed in relying on several shaky assumptions. First among them is that "character"—a term used to describe traits like self-control, perseverance, and curiosity—can be readily dissociated from "cognition." Another supposition is that cognition can be boiled down to that which is measured on an IQ test. And, lastly, Tough's central message assumes that cognitive skills are essentially fixed whereas character is not.

Tough credits this core idea of "character, not cognition" to James Heckman, winner of the 2000 Nobel Prize in Economics. In reanalyzing data from the famous Perry Preschool Project of the 1960s, which provided high-quality preschool education to children living in poverty and then tracked them through adulthood, Heckman found that the factors that best explained the improved life outcomes of the Perry Preschool students were not related to IQ but rather to "noncognitive" factors like curiosity, self-control, and social fluidity. Tough notes that many of the scholars he spoke with while researching this book could be linked to Heckman either directly or indirectly. These scholars include Walter Mischel, who has shown that delay of gratification in preschool is a good predictor of life outcomes,[1,2] and Angela Duckworth, who has demonstrated the importance of "grit" and conscientiousness in achievement.[3, 4]

Tough's tenacity as a journalist is evidenced by the number of scientists he interviewed or learned about in his research for the book. The list reads as a "Who's Who" of developmental and social psychologists, economists, and education researchers. But I found the list short on cognitive scientists, whom I believe would have brought greater clarity to the discussion of the relationships between cognitive abilities and scholastic achievement.

Numerous studies have shown that academic achievement depends on good self-control—a general concept that encompasses layman's terms like "willpower," "character," and "grit," as well as technical terms like "executive functions," "cognitive control processes," or "emotion regulatory strategies." Tough eventually hits on this central concept of self-control and uses it to weave together the different strands of research that he reviewed. Self-control makes it possible to sit still, listen quietly, keep relevant rules and goals in mind, work through feelings rather than erupting in anger, and consider long-term consequences before acting. As such, it makes good sense that research points to self-control as essential for scholastic achievement and good life outcomes.[5]

Self-control can be thought of as a character trait, but it is perhaps more productive to think of it as a set of neurocognitive processes that enable goal-directed thought and behavior. These processes are all instantiated in distributed brain networks that involve the prefrontal cortex, and they are all governed by the same general principles of development and neural plasticity. Considered in this way, it seems unlikely that a child's "character" can change with hard work but that his or her "cognitive abilities" are relatively fixed. A more plausible account is that, with hard work and access to promising interventions, a disadvantaged child can strengthen the neurocognitive skills that enable him or her to excel in school and in life.[6,7]

Quibbles aside, Tough makes a powerful argument for giving children opportunities to rise out of poverty. *How Children Succeed* is likely to bring solace and inspiration to educators—especially those who have lost hope that they can make a difference.

14

To Hell and Back

Brain on Fire: My Month of Madness

By Susannah Cahalan
Reviewed by David Lynch, M.D., Ph.D.

David Lynch, M.D., Ph.D., is a professor of neurology and pediatrics at the University of Pennsylvania, and attending neurologist at the Children's Hospital of Philadelphia. He received his M.D. and Ph.D., specializing in neuropharmacology, from Johns Hopkins University School of Medicine. He was trained in neurology at the University of Pennsylvania, where he has remained for the last 25 years and published more than 100 peer-reviewed papers. More than 40 of these papers are on NMDA receptors, and he is a member of the team at the University of Pennsylvania that initially characterized anti-NMDA-receptor encephalitis.

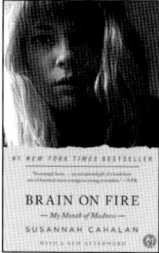

SUPPOSE YOU THOUGHT that you were losing your identity—changing almost overnight and doing things you could not explain. This inexplicable scenario helps launch *Brain on Fire: My Month of Madness,* in which Susannah Cahalan chronicles her experience with the rare disorder anti-NMDA-receptor encephalitis. The onset of this newly identified entity results from the production of antibodies against the N-methyl-D-aspartate (NMDA). The NMDA receptor regulates synaptic plasticity in the brain and is critical for learning and memory. The syndrome evolves over days to weeks and presents as a psychosis similar to that seen in schizophrenia. Cahalan, who suffered from seizures and other neurologic symptoms, creates a medical detective story from a patient's perspective as she captivatingly describes her descent into madness. The result is entertaining reading that gets us to think about the biological origins of our own personalities.

Cahalan presents her experience as a remembrance three years after the onset of her disorder. At the book's beginning, she is a young reporter struggling to find story ideas that will resonate with her editors at the *New York Post.* Over several weeks, she gradually finds her personality changing as she begins suffering from paranoia and hallucinations and withdraws from her friends and colleagues. She describes her thoughts (as she perceives them at the time) and confusion about the origin of her symptoms. Associated with her uncertainty is the realization that the medical community has no explanation for her rapid change.

Eventually, Cahalan develops the neurological features that lead to a correct diagnosis by a medical team at the NYU Langone Medical Center. After weeks of downward spiraling, the team puts all the facts together and arrives at the answer: anti-NMDA-receptor encephalitis, a multistage autoimmune disease that varies wildly in its presentation. Cahalan slowly recovers after undergoing immunomodulatory therapies (steroids, IVIG treatment, and plasmapheresis) in the hospital for a month, plus six months of outpatient follow-up.

For a neuroscientific audience, Cahalan's story stimulates readers to think about the essence of the brain and the personality on several levels. First is the way that she portrays the early events of her encephalitis. She is never sure whether her bizarre behavior is being driven by a disease or is the result of reactions to stress or other events. These reactions include irrational fears of jealously, emotional lability, and even illusions that she can sense other people's thoughts. Almost all of her friends attribute these behaviors to the demands of her job or to psychological issues, but she later realizes that they are subtle early symptoms of anti-NMDA-receptor encephalitis. To a neuroscientist, Cahalan's attempt to decipher her symptoms represents the core of the discussion about where the neurological features of disease end and those of psychiatry begin. It also shows the overlap between these disciplines. In addition, with treatment, the changes in Cahalan's personality reverse not immediately but over long periods of time. This is consistent with the idea of a resilient, structural basis of memory and personality that is attacked by anti-NMDA-receptor encephalitis but is then slowly rebuilt. One issue, which Cahalan describes, is that anti-NMDA-receptor encephalitis is associated with a significant amnesia for the period of acute illness. Cahalan was able to trigger exquisite firsthand details about the experience by revisiting her family's notes, diaries, and recollections, and she points out that it is difficult to tell where her personal memory ends and others' memories begin. This paradox gives readers fascinating insight into the way that people constantly construct their own perceptions.

Brain on Fire is also very effective as an illustration of the diagnostic odyssey of an individual with an unknown disease. Cahalan's signs seem so clear as she presents them three years later, now that the number of individuals reported with anti-NMDA-receptor encephalitis has risen from several hundred to many thousands and the disorder is no longer a research curiosity. Cahalan was diagnosed and treated only after weeks of progression, after finally being admitted to a specialized neurology service in a tertiary care center—and even at that center, the tale is presented as if only one specialist was familiar with the condition. Viewed from that perspective, Cahalan could be considered lucky even to have received the correct diagnosis. The author does not condemn physicians unaware of the diagnosis, as she recognizes the difficulty in staying abreast of new developments in medicine. The book is dedicated to undiagnosed patients, a group that goes far beyond anti-NMDA-receptor encephalitis. Even in the age of whole-genome sequencing, the medical community lacks much information about even common diseases, much less rare ones. Good news is that the NIH Undiagnosed Diseases Program is currently undergoing revision and expansion, in recognition of the number of disorders that remain unknown and might become recognizable.

In some ways, Cahalan oversimplifies the medical and scientific communities. For example, she essentially attributes her diagnosis and the original discovery of anti-NMDA-receptor encephalitis to single individuals. This presentation fails to recognize the manner in which medical care is delivered by interactive teams and the collaborative way that science moves forward. But the perspective presented provides insight on how people outside the medical and scientific communities view their work, and suggests that there is a need for broader education on the collaborative nature of the scientific and medical fields, particularly as the available time for such collaboration decreases.

Another misperception reflects the increasing incidence of anti-NMDA-receptor encephalitis. As Cahalan notes in her discussions with neurological experts, it is easy to speculate that this disorder might be responsible for many episodes of demonic possession or mental illness throughout history. However, this speculation does not match present medical observations, as untreated anti-NMDA-receptor encephalitis is an almost uniformly debilitating disease; scientists have identified few spontaneous survivors who did not have ICU-level care in the acute period. Thus, it seems unlikely that this specific disorder caused such reversible events throughout history, though an as-yet-unappreciated variant conceivably could have done so.

After reading Cahalan's riveting account, neuroscientists may wonder about the specific molecules that influence memory and personality, while physicians may contemplate where the next new disease will arise.

15

Knitting Perspectives

The Autistic Brain: Thinking Across the Spectrum

By Temple Grandin and Richard Panek
Reviewed by Robert L. Findling, M.D., M.B.A.

After serving as the Rocco L. Motto, M.D., Chair of Child and Adolescent Psychiatry at Case Western Reserve University School of Medicine and director of the Division of Child and Adolescent Psychiatry at University Hospitals Case Medical Center, **Robert Findling, M.D., M.B.A.,** joined Johns Hopkins in 2012 as director of child and adolescent psychiatry and vice chair in the Department of Psychiatry and Behavioral Sciences. He is also the vice president for Psychiatry Services and Research at the Kennedy Krieger Institute in Baltimore. Dr. Findling earned his undergraduate degree at Johns Hopkins University and received his medical degree at the Medical College of Virginia. He completed training in pediatrics, psychiatry, and child and adolescent psychiatry at Mt. Sinai Hospital in New York. He subsequently earned his M.B.A. at a joint program run by the London School of Economics and Political Science, the New York University Stern School of Business, and the HEC Paris.

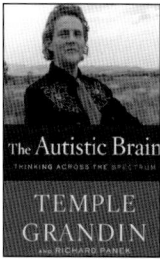

WE DO NOT FULLY UNDERSTAND the neurobiology or the precise causes of autism. While this assertion may be valid, the pace at which researchers are generating scientifically sound information about this syndrome has increased substantively in recent years. *The Autistic Brain,* co-authored by Temple Grandin and science author Richard Panek, is a compelling work that effectively communicates and interweaves several perspectives on autistic disorder that might otherwise seem divergent.

As many readers familiar with autism are aware, Grandin received a Doctor of Philosophy in animal sciences, is a professor of animal science at Colorado State University, and has written and spoken extensively about her own experiences as a person with autistic disorder. Perhaps the world's most famous autistic person, the 62-year-old, cowboy-shirt-wearing Grandin is an animal scientist and slaughterhouse designer who has helped people break through the barriers separating autistic from nonautistic experience.

Grandin's fourth book is both a cogent summary of scientific advances in the field and a personal, firsthand account of the clinical relevance of such research. This makes the book truly translational at its core. In the first chapter, the authors use Grandin's own clinical care as a child as a heuristic reference to how autism has been conceptualized over the years. The authors then engage in a cogent discussion of whether autism could be considered a form of brain damage or a psychological outcome of poor parenting, from both a historical standpoint and a personal perspective. The chapter also contains thoughtful analysis about the impact that nosology has had on people with certain features, such as extreme social anxiety or an autism-spectrum diagnosis. Although the authors are not timid about rendering their own opinions, they effectively communicate their rationales by extrapolating scientific evidence.

Other chapters focus on recent advances in neuroimaging and genetics. The chapter on neuroimaging provides a thoughtful and succinct overview of both the promise and the limitations of this research tool in a way that readers without a neuroscience background can understand. Using, in part, Grandin's techniques from her teaching as a university faculty member, the authors provide a brief primer of the anatomy of the brain, followed by a discussion about the relationship between the clinical correlates of structural brain imaging and behavior. The authors also consider the relevance of functional imaging and studies of brain connectivity to the study of autism, as well as genetics research and its relevance to people with autism.

One particularly interesting chapter focuses on sensory problems in people with autism. According to the authors, sensory disorders appear to be quite common in people with autistic disorder. Grandin notes that some people are oversensitive to stimuli, not responsive enough, or likely to seek out certain types of sensory experiences. Despite the fact that these phenomena are both prevalent and highly distressing to people who experience them, the scientific literature on the topic is surprisingly scant. The same chapter also addresses processing problems associated with the five senses. One of the unique aspects of the book is that its authors suggest means by which to identify people with specific sensory-processing difficulties. In addition, it includes helpful suggestions about how to lessen the burden of people who have these problems.

The authors then grapple with the shortcomings of our descriptive nosology, as well as the putative impact of changes associated with going from DSM-IV to DSM-V (codes that mental-health professionals use to describe the features of mental disorders). Grandin and Panek note that autism is associated with behavioral features in addition to interpersonal ones, albeit oftentimes to a greater degree than in people without autism. Referring to focusing on a single descriptor outside of the context of an individual's unique circumstances as "label locked" thinking, the authors make a compelling argument about how the scientific study of single symptoms, rather than on broader and less precise diagnostic entities, could inspire others to view people with autism as individuals.

The next chapter considers putative strengths of some people with autism. Perhaps more important, the authors discuss how these strengths might manifest themselves in real-world settings. One such strength is attention to detail; another asset is the ability to identify patterns that others do not readily discern. What follows is a chapter that considers putative means by which people with autism might be more successful in mainstream society. The authors do not minimize the vicissitudes of people with autism. Rather, they focus on assets upon which successes can be built.

Due to Grandin's personal accounts of her experience with autism, *The Autistic Brain* is very thoughtfully written. It discusses recent scientific advances about autism in an easily understandable format. At the same time, the authors do a good job of not oversimplifying the material. They pose thoughtful questions about why key clinical features of autistic disorder have not received the amount of scientific research they deserve. Readers will come away from this book appreciating the challenges associated with scientific pursuits while realizing where gaps in research exist. But what makes this book unique and compelling are its firsthand accounts of how autism impacts a person's life, its explanations of why scientific advances matter, and its consideration of the potential strengths of people with autism—despite their difficulties.

Endnotes

1
The Evolution of Risk-Taking

1. Both, C., N. J. Dingemanse, P. J. Drent, and J. M. Tinbergen. 2005. Pairs of extreme avian personalities have highest reproductive success. *Journal of Animal Ecology* 74:667-674.

2. Dugatkin, L. A. 1992. Tendency to inspect predators predicts mortality risk in the guppy, Poecilia reticulata. *Behav. Ecol.* 3:124-128.

3. Dugatkin, L. A. 2009. *Principles of Animal Behavior.* W. W. Norton, New York.

4. Dugatkin, L. A., and M. Alfieri. 2003. Boldness, behavioral inhibition and learning: an evolutionary approach. *Ethology, Ecology & Evolution* 15:43-49.

5. Godin, J. G. J., and L. A. Dugatkin. 1996. Female mating preference for bold males in the guppy, *Poecilia reticulata.* Proceedings of the National Academy of Sciences of the United States of America 93:10262-10267.

6. Kagan, J. 1994. *Galen's Prophecy: Temperament in Human Nature.* Basic Books, New York.

7. Marchetti, C., and P. J. Drent. 2000. Individual differences in the use of social information in foraging by captive great tits. *Anim. Behav.* 60:131-140.

8. Mather, J. A. 2008. To boldly go where no mollusc has gone before: Personality, play, thinking, and consciousness in cephalopods. *American Malacological Bulletin* 24:51-58.

9. Naguib, M., C. Florcke, and K. van Oers. 2011. Effects of Social Conditions During Early Development on Stress Response and Personality Traits in Great Tits (Parus major). *Developmental Psychobiology* 53:592-600.

10. Sinn, D. L., N. A. Moltschaniwskyj, E. Wapstra, and S. R. X. Dall. 2010. Are behavioral syndromes invariant? Spatiotemporal variation in shy/bold behavior in squid. *Behavioral Ecology and Sociobiology* 64:693-702.

11. Weiss, A., M. J. Adams, A. Widdig, and M. S. Gerald. 2011. Rhesus Macaques (Macaca mulatta) as Living Fossils of Hominoid Personality and Subjective Well-Being. *Journal of Comparative Psychology* 125:72-83.

12. Weiss, A., J. E. King, and L. Perkins. 2006. Personality and subjective well-being in orangutans (Pongo pygmaeus and Pongo abelii). *Journal of Personality and Social Psychology* 90:501-511.

13. Wilson, D., K. Coleman, A. Clark, and L. Biederman. 1993. Shy-Bold Continuum in Pumpkinseed Sunfish (*Lepomis gibbosus*): An Ecological Study of a Psychological Trait. *Journal of Comparative Psychology* 107:250-260.

14. Wilson, D. S., A. B. Clark, K. Coleman, and T. Dearstyne. 1994. Shyness and boldness in humans and other animals. *Trends in Ecol. & Evol.* 9:442-446.

2
Hit Parade: The Future of the Sports Concussion Crisis

1. McKee, A.C., Cantu, R.C., Nowinski, C.J., et al. (2009). Chronic traumatic encephalopathy in athletes: Progressive tauopathy following repetitive concussion. *Journal of Neuropathology and Experimental Neurology,* 68, 709-735.

2. Koerte, I.K., Ertl-Wagner, B., Reiser, M., Zafonte, R., Shenton, M.E. (2012). White matter integrity in the brains of professional soccer players without a symptomatic concussion. *JAMA,* 30, 1859-1861.

3. McAllister, T.W., Flashman, L.A., Maerlender, A., et al. (2012). Cognitive effects of one xseason of head impacts in a cohort of collegiate contact sport athletes. *Neurology,* 78, 1777-1784.

4. Blennow, K., Hardy, J., Zetterberg, H. (2012). The Neuropathology and Neurobiology of Traumatic Brain Injury. *Neuron,* 76(5), 886-899.

5. McKee, A.C., Stern, R.A., Nowinski, C.J., et al. (2012). The spectrum of disease in chronic traumatic encephalopathy. Brain. De[1] Arbogast, K.B. et. al. c 2 Epub ahead of print.

6. Cantu, R.C. (2012). *Concussions and Our Kids.* Houghton Mifflin.

7. Arbogast, K.B. Durbin DR, Kallan MJ et al. Injury risk to restrained children exposed to deployed first- and second-generation air bags in frontal crashes. *Arch Pediatr Adolesc Med.* 2005 Apr;15 (4):342-6.

8. Daniel R.W., Rowson, S., Duma S.M. (2012). Head impact exposure in youth football. *Annals of Biomedical Engineering,* 40(4), 976-81. Epub Feb 15.

9. American Academy of Pediatrics (2012). High schools with athletic trainers have more diagnosed concussions, fewer overall injuries. *ScienceDaily.* Retrieved January 2, 2013, from http://www.sciencedaily.com /releases/2012/10/121022080649. htm

10. Echlin P.S., Tator, C.H., Cusimano, M.D. (2010). Return to play after an initial or recurrent concussion in a prospective study of physician-observed junior ice hockey concussions: implications for return to play after a concussion. *Neurosurgical Focus,* 29(5), E5.

11. Beckwith J.G., Greenwald, R.M., Chu, J.J. et al. (2012). Head Impact Exposure Sustained by Football Players on Days of Diagnosed. *Medicine & Science in Sports & Exercise.* Epublished ahead of print.

3
Epilepsy's Big Fat Answer

1. *Ketogenic Diets: Treatments for Epilepsy and Other Disorders* Fifth Edition. Kossoff EH, Freeman JM, et al. demosHealth, New York, 2011.

2. Further information about the ketogenic diet may be obtained from the Charlie Foundation to Help Cure Pediatric Epilepsy.

3. Freeman JM, Vining EPG and Pillas DJ: *Seizures and Epilepsy in Childhood: A Guide for Parents.* Johns Hopkins University Press. Baltimore, Maryland. 3rd edition 2002.

4. Neal EG, Chaffe H, Cross JH, et al. The Ketogenic Diet for the Treatment of Childhood Epilepsy: a Randomized Controlled Trial. *Lancet Neurol.* 2008 Jun; 7(6):500-6

5. Payne NE, Cross JH, Sander JW. The Ketogenic diet and related diets in adoles-

cents and adults: A Review. *Epilepsia,* 52:1941-1948, 2011

6. Freeman JM, Kossoff EH. Ketosis and the Ketogenic Diet 2010: Advances in treating Epilepsy and Other Disorders. *Advances in Pediatrics 2010* 57; 315-329

7. Hartman AL, Rubenstein JE, Kossoff EH. Intermittent fasting: A "new" historical strategy for controlling seizures? Epilepsy Research, J Epilepsy Res. 2012.10.011

8. Seyfried BT, Kiebish M, Marsh J, Mukherjee P. Targeting energy metabolism in brain cancer through calorie restriction and the ketogenic diet. *J Can Res Ther* 2009;5:7-15.

9. Baranano, KW and Hartman, AL The Ketogenic Diet: Uses in Epilepsy and Other Neurologic Illness, *Treat Options Neurol.* 2008;10(6):410.

10. Stafstrom, CE and Rho JM. The Ketogenic Diet as a Treatment Paradigm for Diverse Neurological Disorders Front *Pharmacol.* 2012; 3: 59.

11. Cunnane,S, Nugent S, Roy, M et al. Brain Fuel metabolism, aging and Alzheimer's Disease Nutrition 27 (2011) 3-2

4

Psychiatric Drug Development: Diagnosing a Crisis

1. G. Miller, Is pharma running out of brainy ideas? Science 2010; 329, 502-504; A. Abbott, Novartis to shut brain research facility. *Nature* 2011; 480, 161-162 (2011).

2. World Health Organization. (2008) The Global Burden of Disease. 2004 Update. Geneva: World Health Organization.

3. Munos, B.H., Pharmaceutical Innovation Gets a Little Help from New Friends. *Sci. Transl. Med.* 2013; 5, 168ed1.

4. Hyman, SE. Revolution stalled. *Sci. Transl. Med.* 2012; 4, 155cm11.

5. Nestler, E.J. and Hyman, S. E, Animal models of neuropsychiatric disorders. *Nature Neurosci.* 2010; 13,1161-169.

6. http://www.adni-info.org

7. P. F. Sullivan, Daly, M. J., O'Donovan, M. Genetic architecture of psychiatric disorders: the emerging picture and its implications. *Nature Rev. Genet.* 13, 537-551 (2012).

6

Inside the Letterbox: How Literacy Transforms the Human Brain

1. Nakamura, Kuo, Pegado, Cohen, Tzeng, & Dehaene, 2012

2. Bolger, Perfetti, & Schneider, 2005

3. Dehaene, 2009

4. Josef Parvizi and his team recently discovered that Arabic numerals are recognized by another nearby region. See Shum, Hermes, Foster, Dastjerdi, Rangarajan, Winawer, Miller, & Parvizi, 2013

5. Dehaene, 2005 ; Dehaene & Cohen, 2007

6. Dehaene-Lambertz, Dehaene, & Hertz-Pannier, 2002 ; Dehaene-Lambertz, Hertz-Pannier, Dubois, Meriaux, Roche, Sigman, & Dehaene, 2006

7. Cai, Lavidor, Brysbaert, Paulignan, & Nazir, 2008 ; Pinel & Dehaene, 2009 ; Cai, Paulignan, Brysbaert, Ibarrola, & Nazir, 2010

8. Hasson, Levy, Behrmann, Hendler, & Malach, 2002

9. Biederman, 1987 ; Szwed, Cohen, Qiao, & Dehaene, 2009 ; Szwed, Dehaene,

Kleinschmidt, Eger, Valabregue, Amadon, & Cohen, 2011

10. Tanaka, 1996 ; Baker, Behrmann, & Olson, 2002 ; Brincat & Connor, 2004
11. Dehaene, 2009
12. Changizi, Zhang, Ye, & Shimojo, 2006; See also Changizi & Shimojo, 2005
13. Biederman, 1987 ; Szwed et al., 2011
14. Dehaene, Pegado, Braga, Ventura, Nunes Filho, Jobert, Dehaene-Lambertz, Kolinsky, Morais, & Cohen, 2010
15. Dehaene, Pegado, Braga, Ventura, Nunes Filho, Jobert, Dehaene-Lambertz, Kolinsky, Morais, & Cohen, 2010
16. Szwed, Ventura, Querido, Cohen & Dehaene, 2012
17. For supporting evidence, see also Dundas, Plaut, & Behrmann, 2012 ; Li, Lee, Zhao, Yang, He, & Weng, 2013
18. Monzalvo, Fluss, Billard, Dehaene, & Dehaene-Lambertz, 2012
19. For a review, see chapter 9 in Dehaene, 2009
20. Cornell, 1985
21. Dehaene, Nakamura, Jobert, Kuroki, Ogawa, & Cohen, 2009 ; Pegado, Nakamura, Cohen, & Dehaene, 2011
22. Lachmann & van Leeuwen, 2007
23. Dehaene, Izard, Pica, & Spelke, 2006 ; Kolinsky, Verhaeghe, Fernandes, Mengarda, Grimm-Cabral, & Morais, 2010
24. Jacquemot, Pallier, LeBihan, Dehaene, & Dupoux, 2003
25. Morais, Cary, Alegria, & Bertelson, 1979
26. Ehri & Wilce, 1980 ; Stuart, 1990 ; Ziegler & Ferrand, 1998
27. Thiebaut de Schotten, Cohen, Amemiya, Braga, & Dehaene, 2012
28. National Institute of Child Health and Human Development, 2000
29. Brem, Bach, Kucian, Guttorm, Martin, Lyytinen, Brandeis, & Richardson, 2010

7
Gut Feelings: Bacteria and the Brain

1. Barden, N. (2004). Implication of the hypothalamic-pituitary-adrenal axis in the physiopathology of depression. *J Psychiatry Neurosci* 29, 185-193.
2. Beaumont, W. (1833). Experiments and observations on the gastric juice and the physiology of digestion (Plattsburg: F.P. Allen).
3. Bech-Nielsen, G.V., Hansen, C.H., Hufeldt, M.R., Nielsen, D.S., Aasted, B., Vogensen, F.K., Midtvedt, T., and Hansen, A.K. (2011). Manipulation of the gut microbiota in C57BL/6 mice changes glucose tolerance without affecting weight development and gut mucosal immunity. *Res Vet Sci.*
4. Benton, D., Williams, C., and Brown, A. (2007). Impact of consuming a milk drink containing a probiotic on mood and cognition. *European journal of clinical nutrition* 61, 355-361.
5. Bercik, P., Denou, E., Collins, J., Jackson, W., Lu, J., Jury, J., Deng, Y., Blennerhassett, P., Macri, J., McCoy, K.D., et al. (2011a). The intestinal microbiota affect central levels of brain-derived neurotropic factor and behavior in mice. *Gastroenterology* 141, 599-609, 609 e591-593.
6. Bercik, P., Park, A.J., Sinclair, D., Khoshdel, A., Lu, J., Huang, X., Deng, Y., Blennerhassett, P.A., Fahnestock, M., Moine, D., et al. (2011b). The anxiolytic effect

of Bifidobacterium longum NCC3001 involves vagal pathways for gut-brain communication. *Neurogastroenterol Motil* 23, 1132-1139.

7. Bercik, P., Verdu, E.F., Foster, J.A., Macri, J., Potter, M., Huang, X., Malinowski, P., Jackson, W., Blennerhassett, P., Neufeld, K.A., et al. (2010). Chronic gastrointestinal inflammation induces anxiety-like behavior and alters central nervous system biochemistry in mice. *Gastroenterology* 139, 2102-2112 e2101.

8. Bravo, J.A., Forsythe, P., Chew, M.V., Escaravage, E., Savignac, H.M., Dinan, T.G., Bienenstock, J., and Cryan, J.F. (2011). Ingestion of Lactobacillus strain regulates emotional behavior and central GABA receptor expression in a mouse via the vagus nerve. *Proc Natl Acad Sci U S A* 108, 16050-16055.

9. Cannon, W.B. (1909). The influence of emotional states on the functions of the alimentary canal. *Am J Med Sci* 137, 480-487.

10. Clarke, G., Grenham, S., Scully, P., Fitzgerald, P., Moloney, R.D., Shanahan, F., Dinan, T.G., and Cryan, J.F. (2012). The microbiome-gut-brain axis during early life regulates the hippocampal serotonergic system in a sex-dependent manner. *Mol Psychiatry*.

11. Costello, E.K., Lauber, C.L., Hamady, M., Fierer, N., Gordon, J.I., and Knight, R. (2009). Bacterial community variation in human body habitats across space and time. *Science* 326, 1694-1697.

12. Desbonnet, L., Garrett, L., Clarke, G., Kiely, B., Cryan, J.F., and Dinan, T.G. (2010). Effects of the probiotic Bifidobacterium infantis in the maternal separation model of depression. *Neuroscience* 170, 1179-1188.

13. Foster, J.A., and McVey Neufeld, K.A. (2013). Gut-brain axis: how the microbiome influences anxiety and depression. *Trends Neurosci* 36, 305-312.

14. Furness, J.B. (2012). The enteric nervous system and neurogastroenterology. Nature reviews Gastroenterology & hepatology 9, 286-294.

15. Gareau, M.G., Jury, J., MacQueen, G., Sherman, P.M., and Perdue, M.H. (2007). Probiotic treatment of rat pups normalises corticosterone release and ameliorates colonic dysfunction induced by maternal separation. *Gut* 56, 1522-1528.

16. Gareau, M.G., Silva, M.A., and Perdue, M.H. (2008). Pathophysiological mechanisms of stress-induced intestinal damage. *Current molecular medicine* 8, 274-281.

17. Gill, S.R., Pop, M., Deboy, R.T., Eckburg, P.B., Turnbaugh, P.J., Samuel, B.S., Gordon, J.I., Relman, D.A., Fraser-Liggett, C.M., and Nelson, K.E. (2006). Metagenomic analysis of the human distal gut microbiome. *Science* 312, 1355-1359.

18. Goehler, L.E., Park, S.M., Opitz, N., Lyte, M., and Gaykema, R.P. (2008). Campylobacter jejuni infection increases anxiety-like behavior in the holeboard: possible anatomical substrates for viscerosensory modulation of exploratory behavior. *Brain Behav Immun* 22, 354-366.

19. Heijtz, R.D., Wang, S., Anuar, F., Qian, Y., Bjorkholm, B., Samuelsson, A., Hibberd, M.L., Forssberg, H., and Pettersson, S. (2011). Normal gut microbiota modulates brain development and behavior. *Proc Natl Acad Sci U S A* 108, 3047-3052.

20. Hooper, L.V., Wong, M.H., Thelin, A., Hansson, L., Falk, P.G., and Gordon, J.I. (2001). Molecular analysis of commensal host-microbial relationships in the intestine. *Science* 291, 881-884.

21. Jumpertz, R., Le, D.S., Turnbaugh, P.J., Trinidad, C., Bogardus, C., Gordon, J.I., and Krakoff, J. (2011). Energy-balance studies reveal associations between gut

microbes, caloric load, and nutrient absorption in humans. *Am J Clin Nutr* 94, 58-65.

22. Kau, A.L., Ahern, P.P., Griffin, N.W., Goodman, A.L., and Gordon, J.I. (2011). Human nutrition, the gut microbiome and the immune system. *Nature* 474, 327-336.

23. Lyte, M., Li, W., Opitz, N., Gaykema, R.P., and Goehler, L.E. (2006). Induction of anxiety-like behavior in mice during the initial stages of infection with the agent of murine colonic hyperplasia Citrobacter rodentium. *Physiol Behav* 89, 350-357.

24. Macpherson, A.J., and Harris, N.L. (2004). Interactions between commensal intestinal bacteria and the immune system. *Nat Rev Immunol* 4, 478-485.

25. Messaoudi, M., Violle, N., Bisson, J.F., Desor, D., Javelot, H., and Rougeot, C. (2011). Beneficial psychological effects of a probiotic formulation (Lactobacillus helveticus R0052 and Bifidobacterium longum R0175) in healthy human volunteers. *Gut microbes* 2, 256-261.

26. Neufeld, K.A., Kang, N., Bienenstock, J., and Foster, J.A. (2011a). Effects of intestinal microbiota on anxiety-like behavior. *Commun Integr Biol* 4, 492-494.

27. Neufeld, K.M., Kang, N., Bienenstock, J., and Foster, J.A. (2011b). Reduced anxiety-like behavior and central neurochemical change in germ-free mice. *Neurogastroenterol Motil* 23, 255-264, e119.

28. Pavlov, I. (1910). *The work of digestive glands.* [English translation from Russian by W. H. Thompson.] (London: Griffen).

29. Qin, J., Li, R., Raes, J., Arumugam, M., Burgdorf, K.S., Manichanh, C., Nielsen, T., Pons, N., Levenez, F., Yamada, T., et al. (2010). A human gut microbial gene catalogue established by metagenomic sequencing. *Nature* 464, 59-65.

30. Rao, A.V., Bested, A.C., Beaulne, T.M., Katzman, M.A., Iorio, C., Berardi, J.M., and Logan, A.C. (2009). A randomized, double-blind, placebo-controlled pilot study of a probiotic in emotional symptoms of chronic fatigue syndrome. *Gut Pathog* 1, 6.

31. Schloissnig, S., Arumugam, M., Sunagawa, S., Mitreva, M., Tap, J., Zhu, A., Waller, A., Mende, D.R., Kultima, J.R., Martin, J., et al. (2013). Genomic variation landscape of the human gut microbiome. *Nature* 493, 45-50.

32. Serino, M., Chabo, C., and Burcelin, R. (2012). Intestinal MicrobiOMICS to define health and disease in human and mice. *Current pharmaceutical biotechnology* 13, 746-758.

33. Sudo, N., Chida, Y., Aiba, Y., Sonoda, J., Oyama, N., Yu, X.N., Kubo, C., and Koga, Y. (2004). Postnatal microbial colonization programs the hypothalamic-pituitary-adrenal system for stress response in mice. *J Physiol* 558, 263-275.

34. Teitelbaum, A.A., Gareau, M.G., Jury, J., Yang, P.C., and Perdue, M.H. (2008). Chronic peripheral administration of corticotropin-releasing factor causes colonic barrier dysfunction similar to psychological stress. *Am J Physiol Gastrointest Liver Physiol* 295, G452-459.

8
Do Cytokines Really Sing the Blues?

1. Miller, A. H., Maletic V., Raison C. L. (2009). Inflammation and its discontents:

the role of cytokines in the pathophysiology of major depression. *Biol Psychiatry* 65, 732-741.

2. Vaudo, G., Marchesi S., Gerli R., Allegrucci R., Giordano A., Siepi D., Pirro M., Shoenfeld Y., Schillaci G., Mannarino E. (2004). Endothelial dysfunction in young patients with rheumatoid arthritis and low disease activity. *Ann Rheum Dis* 63, 31-35.

3. Chamouard, P., Richert Z., Meyer N., Rahmi G., Baumann R. (2006). Diagnostic value of C-reactive protein for predicting activity level of Crohn's disease. *Clin Gastroenterol Hepatol* 4, 882-887.

4. Hoeboer, S. H., Groeneveld A. B. (2013). Changes in circulating procalcitonin versus C-reactive protein in predicting evolution of infectious disease in febrile, critically ill patients. *PLoS One* 8, e65564.

5. Wium-Andersen, M. K., Orsted D. D., Nielsen S. F., Nordestgaard B. G. (2012). Elevated C-Reactive Protein Levels, Psychological Distress, and Depression in 73 131 Individuals. *Arch Gen Psychiatry*, 1-9.

6. Musselman, D. L., Miller A. H., Porter M. R., Manatunga A., Gao F., Penna S., Pearce B. D., Landry J., Glover S., McDaniel J. S., Nemeroff C. B. (2001). Higher than normal plasma interleukin-6 concentrations in cancer patients with depression: preliminary findings. *Am J Psychiatry* 158, 1252-1257.

7. Couzin-Frankel, J. (2010). Inflammation bares a dark side. *Science* 330, 1621.

8. Raison, C. L., Rutherford R. E., Woolwine B. J., Shuo C., Schettler P., Drake D. F., Haroon E., Miller A. H. (2013). A randomized controlled trial of the tumor necrosis factor antagonist infliximab for treatment-resistant depression: the role of baseline inflammatory biomarkers. *JAMA Psychiatry* 70, 31-41.

9. Yirmiya, R., Goshen I. (2011). Immune modulation of learning, memory, neural plasticity and neurogenesis. *Brain Behav Immun* 25, 181-213.

10. Perez-Caballero, L., Perez-Egea R., Romero-Grimaldi C., Puigdemont D., Molet J., Caso J. R., Mico J. A., Perez V., Leza J. C., Berrocoso E. (2013). Early responses to deep brain stimulation in depression are modulated by anti-inflammatory drugs. *Mol Psychiatry*.

11. Warner-Schmidt, J. L., Vanover K. E., Chen E. Y., Marshall J. J., Greengard P. (2011). Antidepressant effects of selective serotonin reuptake inhibitors (SSRIs) are attenuated by antiinflammatory drugs in mice and humans. *Proc Natl Acad Sci U S A* 108, 9262-9267.

12. Raison, C. L., Lowry C. A., Rook G. A. (2010). Inflammation, sanitation, and consternation: loss of contact with coevolved, tolerogenic microorganisms and the pathophysiology and treatment of major depression. *Arch Gen Psychiatry* 67, 1211-1224.

13. Raison, C. L., Miller A. H. (2013). The evolutionary significance of depression in Pathogen Host Defense (PATHOS-D). *Mol Psychiatry* 18, 15-37.

14. Capuron, L., Miller A. H. (2004). Cytokines and psychopathology: lessons from interferon-alpha. *Biol Psychiatry* 56, 819-824.

15. Miller, A. H., Haroon E., Raison C. L., Felger J. C. (2013). Cytokine targets in the brain: impact on neurotransmitters and neurocircuits. *Depress Anxiety* 30, 297-306.

16. Harrison, N. A., Brydon L., Walker C., Gray M. A., Steptoe A., Critchley H. D. (2009). Inflammation causes mood changes through alterations in subgenual cingulate activity and mesolimbic connectivity. *Biol Psychiatry* 66, 407-414.

17. Slavich, G. M., Way B. M., Eisenberger N. I., Taylor S. E. (2010). Neural sensitivity to social rejection is associated with inflammatory responses to social stress. *Proc Natl Acad Sci U S A* 107, 14817-14822.
18. Capuron, L., Pagnoni G., Drake D. F., Woolwine B. J., Spivey J. R., Crowe R. J., Votaw J. R., Goodman M. M., Miller A. H. (2012). Dopaminergic mechanisms of reduced basal ganglia responses to hedonic reward during interferon alfa administration. *Arch Gen Psychiatry* 69, 1044-1053.
19. Felger, J. C., Mun J., Kimmel H. L., Nye J. A., Drake D. F., Hernandez C. R., Freeman A. A., Rye D. B., Goodman M. M., Howell L. L., Miller A. H. (2013). Chronic Interferon-alpha Decreases Dopamine 2 Receptor Binding and Striatal Dopamine Release in Association with Anhedonia-like Behavior in Non-Human Primates. *Neuropsychopharmacology.*
20. Koo, J. W., Duman R. S. (2008). IL-1beta is an essential mediator of the antineurogenic and anhedonic effects of stress. *Proc Natl Acad Sci U S A* 105, 751-756.
21. Pariante, C. M., Miller A. H. (2001). Glucocorticoid receptors in major depression: relevance to pathophysiology and treatment. *Biol Psychiatry* 49, 391-404.
22. Raison, C. L., Borisov A. S., Woolwine B. J., Massung B., Vogt G., Miller A. H. (2010). Interferon-alpha effects on diurnal hypothalamic-pituitary-adrenal axis activity: relationship with proinflammatory cytokines and behavior. *Mol Psychiatry* 15, 535-547.
23. Frank, M. G., Baratta M. V., Sprunger D. B., Watkins L. R., Maier S. F. (2007). Microglia serve as a neuroimmune substrate for stress-induced potentiation of CNS pro-inflammatory cytokine responses. *Brain Behav Immun* 21, 47-59.
24. Steiner, J., Bielau H., Brisch R., Danos P., Ullrich O., Mawrin C., Bernstein H. G., Bogerts B. (2008). Immunological aspects in the neurobiology of suicide: elevated microglial density in schizophrenia and depression is associated with suicide. *J Psychiatr Res* 42, 151-157.
25. Pace, T. W., Mletzko T. C., Alagbe O., Musselman D. L., Nemeroff C. B., Miller A. H., Heim C. M. (2006). Increased stress-induced inflammatory responses in male patients with major depression and increased early life stress. *Am J Psychiatry* 163, 1630-1633.

9
ADHD: 10 Years Later

1. Almeida, L. G., J. Ricardo-Garcell, et al. (2011). "Reduced right frontal cortical thickness in children, adolescents and adults with ADHD and its correlation to clinical variables: A cross-sectional study." *Journal of Psychiatric Research* 44(16): 1214-1223.
2. Batty, M. J., E. B. Liddle, et al. (2010). "Cortical Gray Matter in Attention-Deficit/Hyperactivity Disorder: A Structural Magnetic Resonance Imaging Study." *Journal of the American Academy of Child & Adolescent Psychiatry* 49(3): 229-238.
3. Bledsoe, J., M. Semrud-Clikeman, et al. (2009). "A Magnetic Resonance Imaging Study of the Cerebellar Vermis in Chronically Treated and Treatment-Naïve Children with Attention Deficit/Hyperactivity Disorder Combined Type." *Biological Psychiatry.*
4. Bressler, S. L. and V. Menon (2010). "Large-scale brain networks in cognition: emerging methods and principles." *Trends in Cognitive Sciences* 14(6): 277-290.

5. Castellanos, F., P. Lee, et al. (2002). "Developmental trajectories of brain volume abnormalities in children and adolescents with attention-deficit/hyperactivity disorder." *Jama* 288: 1740 - 1748.

6. Castellanos, F. X. and R. Tannock (2002). "Neuroscience of attention-deficit/hyperactivity disorder: the search for endophenotypes." *Nature Reviews Neuroscience* 3(8): 617-628.

7. Cramer, S. C., M. Sur, et al. (2011). "Harnessing neuroplasticity for clinical applications." *Brain* 134(6): 1591-1609.

8. Daley, D. and J. Birchwood (2010). "ADHD and academic performance: why does ADHD impact on academic performance and what can be done to support ADHD children in the classroom?" *Child: Care, Health and Development* 36(4): 455-464.

9. Durston, S., H. Hulshoff Pol, et al. (2004). "Magnetic resonance imaging of boys with attention-deficit/hyperactivity disorder and their unaffected siblings." *Journal of the American Academy of Child & Adolescent Psychiatry* 43: 332 - 340.

10. Faraone, S. V., J. Biederman, et al. (2006). "The age-dependent decline of attention deficit hyperactivity disorder: a meta-analysis of follow-up studies." *Psychological Medicine* 36(2): 159-165.

11. Harpin, V. A. (2005). "The effect of ADHD on the life of an individual, their family, and community from preschool to adult life." *Archives of disease in childhood* 90(suppl 1): i2-i7.

12. Klingberg, T. (2010). "Training and plasticity of working memory." *Trends in Neurosciences* 14(7): 317.

13. Konrad, K. and S. B. Eickhoff (2010). "Is the ADHD brain wired differently? A review on structural and functional connectivity in attention deficit hyperactivity disorder." *Human Brain Mapping* 31(6): 904-916.

14. Kraemer, H. C., J. A. Yesavage, et al. (2000). "How Can We Learn About Developmental Processes From Cross-Sectional Studies, or Can We?" *Am J Psychiatry* 157(2): 163-171.

15. Lawrence, K. E., J. G. Levitt, et al. (2013). "White Matter Microstructure in Attention-Deficit/Hyperactivity Disorder Subjects and Their Siblings." *Journal of the American Academy of Child & Adolescent Psychiatry.*

16. Mackie, S., P. Shaw, et al. (2007). "Cerebellar development and clinical outcome in attention deficit hyperactivity disorder.[see comment]." *American Journal of Psychiatry* 164(4): 647-655.

17. Narr, K. L., R. P. Woods, et al. (2009). "Widespread Cortical Thinning Is a Robust Anatomical Marker for Attention-Deficit/Hyperactivity Disorder." *Journal of the American Academy of Child & Adolescent Psychiatry* 48(10): 1014-1022.

18. Nigg, J. T. and B. J. Casey (2005). "An integrative theory of attention-deficit/hyperactivity disorder based on the cognitive and affective neurosciences." *Development & Psychopathology* 17(3): 785-806.

19. Petersen, S. E. and M. I. Posner (2012). "The Attention System of the Human Brain: 20 Years After." *Annual Review of Neuroscience* 35(1): 73-89.

20. Posner, M. I. and S. E. Petersen (1990). "The attention system of the human brain." *Annual Review of Neuroscience* 13: 25-42.

21. Rubia, K., J. Noorloos, et al. (2003). "Motor timing deficits in community and clinical boys with hyperactive behavior: the effect of methylphenidate on motor timing." *Journal of Abnormal Child Psychology* 31(3): 301-313.

22. Rubia, K., A. Taylor, et al. (1999). "Synchronization, anticipation, and consistency in motor timing of children with dimensionally defined attention deficit hyperactivity behaviour." *Perceptual & Motor Skills* 89(3 Pt 2): 1237-1258.

23. Shaw, P., J. Lerch, et al. (2006). "Longitudinal mapping of cortical thickness and clinical outcome in children and adolescents with attention-deficit/hyperactivity disorder." *Archives of General Psychiatry* 63(5): 540-549.

24. Shaw, P., M. Malek, et al. (2013). "Trajectories of Cerebral Cortical Development in Childhood and Adolescence and Adult Attention-Deficit/Hyperactivity Disorder." *Biological Psychiatry.*

25. Smith, A., E. Taylor, et al. (2002). "Evidence for a pure time perception deficit in children with ADHD." *Journal of Child Psychology & Psychiatry & Allied Disciplines* 43(4): 529-542.

26. Sobel, L. J., R. Bansal, et al. (2010). "Basal ganglia surface morphology and the effects of stimulant medications in youth with attention deficit hyperactivity disorder." *American Journal of Psychiatry* 167(8): 977-986.

27. Sonuga-Barke, E. J., T. Saxton, et al. (1998). "The role of interval underestimation in hyperactive children's failure to suppress responses over time." *Behavioural Brain Research* 94(1): 45-50.

28. van Ewijk, H., D. J. Heslenfeld, et al. (2012). "Diffusion tensor imaging in attention deficit/hyperactivity disorder: A systematic review and meta-analysis." *Neuroscience & Biobehavioral Reviews* 36(4): 1093-1106.

29. Volkow, N. D., G.-J. Wang, et al. (2009). "Evaluating dopamine reward pathway in ADHD: clinical implications.[Erratum appears in JAMA. 2009 Oct 7;302(13):1420]." *JAMA* 302(10): 1084-1091.

30. Wehmeier, P. M., A. Schacht, et al. (2010). "Social and Emotional Impairment in Children and Adolescents with ADHD and the Impact on Quality of Life." *Journal of Adolescent Health* 46(3): 209-217.

10
Lewy Body Dementia: The Under-Recognized But Common Foe

1. http://www.lbda.org/category/3437/what-is-lbd.htm
2. McKeith IG, Dickson DW, Lowe J, et al; Consortium on DLB. Diagnosis and management of dementia with Lewy bodies: third report of the DLB Consortium. *Neurology* 65:1863–1872 (2005).
3. Galvin JE, Pollack J Morris JC. Clinical phenotype of Parkinson Disease Dementia. *Neurology* 67:1605–1611 (2006).
4. Barker WW, Luis CA, Kashuba A et al. Relative frequencies of Alzheimer Disease, Lewy Body, vascular, frontotemporal dementia, and hippocampal sclerosis in the State of Florida Brain Bank. *Alzhemer. Dis. Assoc. Disord.* 16:203–212 (2012).
5. Galvin JE, Duda JE, Kaufer DI, Lippa CF, Taylor A, Zarit SH. Lewy body dementia: The caregiver experience of clinical care. *Parkinson Rel Disord,* 16:388–392 (2010).
6. Galvin JE, Duda JE, Kaufer DI, Lippa CF, Taylor A, Zarit SH. Lewy body dementia: Caregiver burden and unmet needs. *Alz Dis Assoc Disord,* 24:177–181 (2010).
7. Leggett, AN, Zarit, S, Taylor A, Galvin, JE. Stress and burden among caregivers of patients with Lewy Body Dementia. *Gerontoloist.* 51:76–85 (2011).
8. Ferman TJ, Smith GE, Boeve BF, Ivnik RJ, Petersen RC, Knopman D, Graff-Rad-

ford N, Parisi J, Dickson DW. DLB fluctuations: specific features that reliably differentiate DLB from AD and normal aging. *Neurology.* 62:181–187 (2004).

9. Auning E et al. Early and presenting symptoms of dementia with lewy bodies. *Dement Geriatr Cogn Disord.* 32: 202–208 (2011).

10. Burn DJ. Parkinson's disease dementia: what's in a Lewy body? *J. Neural. Transm. Suppl.* 70:361–365 (2006).

11. Johnson DK, Morris JC, Galvin JE. Verbal and visuospatial deficits in dementia with Lewy bodies. *Neurology.* 65:1232–1238 (2005).

12. Karantzoulis S, Galvin JE. Update on Dementia with Lewy Bodies. *Curr Translat Geriatr Exp Gerontol Rep,* 2:196–204 (2013).

13. Thaipisuttikul P, Lobach I, Zweig Y, Gurnani A, Galvin JE. Capgras Syndrome in Dementia with Lewy Bodies. *Int Psychogeriatr* 25:843–849 (2013).

14. Stavitsky K, Brickman AM, Scarmeas N et al. The progression of cognition, psychiatric symptoms and functional abilities in dementia with Lewy Body and Alzheimer's Disease. *Arch Neurol* 63:1450–1456 (2006).

15. Collerton D et al. Systematic review and meta-analysis show that dementia with Lewy bodies is a visual-perceptual and attentional executive dementia. *Dement Geriatr Cogn Disord.* 16:229–237 (2003).

16. Galvin JE et al. Personality traits distinguishing dementia with Lewy bodies from Alzheimer disease. *Neurology.* 68:1895–1901 (2007).

17. Tawarneh R, Galvin JE. Distinguishing Lewy Body Dementia from Alzheimer's Disease. *Expert Review Neurotherapeutics.* 7:1499–1516 (2007).

18. Weisman D, Cho M, Taylor, C, Adam, A, Thal LJ, Hansen LA. In dementia with Lewy bodies, Braak stage determines phenotypes, not Lewy body distribution. *Neurology,* 69: 356–359 (2007).

19. Sabbagh MN, Corey-Bloom J, Tiraboschi P, Thomas R, Masliah E, Thal LJ. Neurochemical markers do not correlate with cognitive decline in the Lewy body variant of Alzheimer disease. *Arch Neurol,* 56:1458–1461 (1999).

20. Whitwell JL et al. Focal atrophy in dementia with Lewy bodies on MRI: a distinct pattern from Alzheimer's disease. *Brain.* 130:708–719 (2007).

21. O'Brien JT et al. Dopamine transporter loss visualized with FP-CIT SPECT in the differential diagnosis of dementia with Lewy bodies. *Arch Neurol.* 61:919–925 (2004).

22. Walker Z et al. Differentiation of dementia with Lewy bodies from Alzheimer's disease using a dopaminergic presynaptic ligand. *Neurol Neurosurg Psychiatry.* 73:134–140 (2002).

23. McKeith I et al. Neuroleptic sensitivity in patients with senile dementia of Lewy body type. *BMJ.* 305:673–678 (1992).

24. Rogan S, Lippa CF. Alzheimer's disease and other dementias: a review. *Am J Alzheimers Dis Other Demen.* 17:11–17 (2002).

11

Getting High on the Endocannabinoid System

1. Gaoni, Y. and Mechoulam, R. (1964). Isolation, structure and partial synthesis of an active constituent of Hashish. *Journal of the American Chemical Society* 86, 1646–1647.

2. Devane, W.A., Dysarz, F.A. 3rd, Johnson, M.R., Melvin, L.S. and Howlett, A.C. (1988). Determination and characterization of a cannabinoid receptor in rat brain. *Molecular Pharmacology* 34, 605–613.

3. Matsuda, L.A., Lolait, S.J., Brownstein, M.J., Young, A.C. and Bonner, T.I. (1990). Structure of a cannabinoid receptor and functional expression of the cloned cDNA. *Nature* 346, 561–564.

4. Munro, S., Thomas, K.L. and Abu-Shaar, M. (1993). Molecular characterization of a peripheral receptor for cannabinoids. *Nature* 365, 61–65.

5. Devane, W.A., Hanus, L., Breuer, A., Pertwee, R.G., Stevenson, L.A., Griffin, G., Gibson, D., Mandelbaum, A., Etinger, A. and Mechoulam, R. (1992). Isolation and structure of a brain constituent that binds to the cannabinoid receptor. *Science*, 258, 1946–1949.

6. Mechoulam, R., Ben-Shabat, S., Hanus, L., Ligumsky, M., Kaminski, N.E., Schatz, A.R., Gopher, A., Almog, S., Martin, B.R., Compton, D.R., Pertwee, R.G., Griffin, G., http://www.sciencedirect.com/science/article/pii/000629529500109D - AFF5Bayewitch, M., Barg, J. and Vogel, Z. (1995). Identification of an endogenous 2-monoglyceride, present in canine gut, that binds to cannabinoid receptors. *Biochemical Pharmacology* 50, 83–90.

7. Sugiura, T., Kondo, S., Sukagawa, A., Nakane, S., Shinoda, A., Itoh, K., Yamashita, A. and Waku, K. (1995). 2-Arachidonoylglycerol: a possible endogenous cannabinoid receptor ligand in brain. *Biochemical and Biophysical Research Communications* 215, 89–97.

8. Wilson, R.I. and Nicoll, R.A. (2001) Endogenous cannabinoids mediate retrograde signalling at hippocampal synapses. *Nature* 410, 588–592.

9. Ohno-Shosaku, T., Maejima, T. and Kano, M. (2001) Endogenous cannabinoids mediate retrograde signals from depolarized postsynaptic neurons to presynaptic terminals. *Neuron* 29, 729–738.

10. Freund, T.F., Katona, I. and Piomelli, D. (2003). Role of endogenous cannabinoids in synaptic signaling. *Physiological Reviews* 83, 1017–1066.

10. Osei-Hyiaman, D., DePetrillo, M., Pacher, P., Liu, J., Radaeva, S., Batkai, S., Harvey-White, J., Mackie, K., Offertaler, L., Wang, L. and Kunos, G. (2005). Endocannabinoid activation at hepatic CB1 receptors stimulates fatty acid synthesis and contributes to diet-induced obesity. *Journal of Clinical Investigation* 115, 1298–1305.

11. Ravinet Trillou, C., Arnone, M., Delgorge, C., Gonalons, N., Keane, P., Maffrand, J.P. and Soubrie, P. (2003). Anti-obesity effect of SR141716, a CB1 receptor antagonist, in diet-induced obese mice. *American Journal of Physiology. Regulatory, Integrative and Comparative Physiology* 284, R345–R353.

12. Marsicano, G., Wotjak, C.T., Azad, S.C., Bisogno, T., Rammes, G., Cascio, M.G., Hermann, H., Tang, J., Hofmann, C., Zieglgänsberger, W., Di Marzo, V. and Lutz, B. (2002). The endogenous cannabinoid system controls extinction of aversive memories. *Nature* 418, 530–534.

13. Zhang, P.W., Ishiguro, H., Ohtsuki, T., Hess, J., Carillo, F., Walther, D., Onaivi, E.S., Arinami, T. and Uhl, G.R. (2004). Human cannabinoid receptor 1: 5' exons, candidate regulatory regions, polymorphisms, haplotypes and association with polysubstance abuse. *Molecular Psychiatry* 9, 916–931.

14. Filbey, F.M., Schacht, J.P., Myers, U.S., Chavez, R.S. and Hutchison, K.E. (2010).

Individual and additive effects of the CNR1 and FAAH genes on brain response to marijuana cues. *Neuropsychopharmacology* 35, 967–975.

15. Ben-Shabat, S., Fride, E, Sheskin, T., Tamiri, T., Rhee, M.H., Vogel, Z., Bisogno, T., De Petrocellis, L., Di Marzo, V., and Mechoulam, R. (1998). An entourage effect: inactive endogenous fatty acid glycerol esters enhance 2-arachidonoyl-glycerol cannabinoid activity. *European Journal of Pharmacology* 353, 23–31.

16. Pertwee, R.G. (2008). The diverse CB1 and CB2 receptor pharmacology of three plant cannabinoids: delta9-tetrahydrocannabinol, cannabidiol and delta9-tetrahydrocannabivarin. *British Journal of Pharmacology* 153, 199–215.

17. Micale, V., Di Marzo, V., Sulcova, A., Wotjak, C.T. and Drago, F. (2013). Endocannabinoid system and mood disorders: priming a target for new therapies. *Pharmacology and Therapeutics* 13, 18–37.

18. Stella, N. (2009). Endocannabinoid signaling in microglial cells. *Neuropharmacology* 56, Suppl 1:244–253.

19. Young, S., (2013). Marijuana stops child's seizures. http://www.cnn.com/2013/08/07/health/charlotte-child-medical-marijuana/index.html?iref=allsearch

20. Hill, T.M.D., Cascio, M.G., Romano, B., Duncan, M., Pertwee, R.G., Williams, C.M., Whalley, B.J. and Hill, A.J. (2013). Cannabidivarin-rich cannabis extracts are anticonvulsant in mouse and rat via a CB1 receptor-independent mechanism. *British Journal of Pharmacology* 170, 679–692.

21. Gaetani, S., Cuomo, V. and Piomelli, D. (2003). Anandamide hydrolysis: a new target for anti-anxiety drugs? *Trends in Molecular Medicine* 9, 474–478.

22. Long, J.Z., Li, W., Booker, L., Burston, J.J., Kinsey, S.G., Schlosburg, J.E., Pavón, F.J., Serrano, A.M., Selley, D.E., Parsons, L.H., Lichtman, A.H. and Cravatt, BF. (2009) Selective blockade of 2-arachidonoylglycerol hydrolysis produces cannabinoid behavioral effects. *Nature Chemical Biology* 5, 37–44.

23. Kathuria, S., Gaetani, S., Fegley, D., Valiño, F., Duranti, A., Tontini, A., Mor, M., Tarzia, G., La Rana, G., Calignano, A., Giustino, A., Tattoli, M., Palmery, M., Cuomo, V. and Piomelli, D. (2003). Modulation of anxiety through blockade of anandamide hydrolysis. *Nature Medicine* 9, 76–81.

24. Ross, R.A. (2003). Anandamide and vanilloid TRPV1 receptors. *British Journal of Pharmacology* 140, 790–801.

25. Micale, V., Cristino, L., Tamburella, A., Petrosino, S., Leggio, G.M., Drago, F. and Di Marzo, V. (2009). Anxiolytic effects in mice of a dual blocker of fatty acid amide hydrolase and transient receptor potential vanilloid type-1 channels. *Neuropsychopharmacology* 34, 593–606.

26. Schlosburg, J.E., Blankman, J.L., Long, J.Z., Nomura, D.K., Pan, B., Kinsey, S.G., Nguyen, P.T., Ramesh, D., Booker, L., Burston, J.J., Thomas, E.A., Selley, D.E., Sim-Selley, L.J., Liu, Q.S., Lichtman, A.H. and Cravatt, B.F. (2010). Chronic monoacylglycerol lipase blockade causes functional antagonism of the endocannabinoid system. *Nature Neuroscience* 13, 1113–1119.

27. Spradley, J.M., Guindon, J. and Hohmann, A.G. (2010). Inhibitors of monoacylglycerol lipase, fatty-acid amide hydrolase and endocannabinoid transport differentially suppress capsaicin-induced behavioral sensitization through peripheral endocannabinoid mechanisms. *Pharmacological Research* 62, 249–258.

28. Varma, N., Carlson, G.C., Ledent, C. and Alger, B.E. (2001). Metabotropic glu-

tamate receptors drive the endocannabinoid system in hippocampus. *Journal of Neuroscience* 21, RC188, 1–5.

29. Maejima, T., Hashimoto, K., Yoshida, T., Aiba, A. and Kano, M. (2001). Presynaptic inhibition caused by retrograde signal from metabotropic glutamate to cannabinoid receptors. *Neuron* 31, 463–475.

30. Bear, M.F., Huber, K.M. and Warren, S.T. (2004). The mGluR theory of fragile X mental retardation. *Trends in Neuroscience* 27, 370–377.

31. Zhang, L. and Alger, B.E. (2010). Enhanced endocannabinoid signaling elevates neuronal excitability in fragile X syndrome. *Journal of Neuroscience* 30, 5724–5729.

32. Maccarrone, M., Rossi, S., Bari, M., De Chiara, V., Rapino, C., Musella, A., Bernardi, G., Bagni, C. and Centonze, D. (2010). Abnormal mGlu 5 receptor/endocannabinoid coupling in mice lacking FMRP and BC1 RNA. *Neuropsychopharmacology* 35, 1500–1509.

13
Ain't No Mountain High Enough

1. Mischel W, Ayduk O, Berman MG, Casey BJ, Gotlib IH, Jonides J, Kross E, Teslovich T, Wilson NL, Zayas V, Shoda Y. 'Willpower' over the life span: decomposing self-regulation. *Soc Cogn Affect Neurosci*. 2011 Apr;6(2):252-6. doi: 10.1093/scan/nsq081.

2. Casey BJ, Somerville LH, Gotlib IH, Ayduk O, Franklin NT, Askren MK, Jonides J, Berman MG, Wilson NL, Teslovich T, Glover G, Zayas V, Mischel W, Shoda Y. *Proc Natl Acad Sci U S A*. 2011 Sep 6;108(36):14998-5003. doi: 10.1073/pnas.1108561108.

3. Duckworth, A. L. & Seligman, M. E. P. (2005). Self-discipline outdoes IQ predicting academic performance in adolescents. *Psychological Science,* 16, 939-944.

4. Duckworth, A. L., Quinn, P., Tsukayama, E. (2012). What No Child Left Behind leaves behind: The roles of IQ and self-control in predicting standardized achievement test scores and report card grades. *Journal of Educational Psychology,* 104(2), 439-451.

5. Moffitt TE, Arseneault L, Belsky D, Dickson N, Hancox RJ, Harrington H, Houts R, Poulton R, Roberts BW, Ross S, Sears MR, Thomson WM, Caspi A. A gradient of childhood self-control predicts health, wealth, and public safety. *Proc Natl Acad Sci U S A*. 2011 Feb 15;108(7):2693-8. doi: 10.1073/pnas.1010076108.

6. Mackey, A.P., Hill, S.S., Stone, S.I., & Bunge, S.A. (2011) Dissociable effects of reasoning and speed training in children. *Developmental Science,* May;14(3):582-90.

7. Bergman Nutley S, Söderqvist S, Bryde S, Thorell LB, Humphreys K, Klingberg T. Gains in fluid intelligence after training non-verbal reasoning in 4-year-old children: a controlled, randomized study. *Dev Sci*. 2011 May;14(3):591-601. doi: 10.1111/j.1467-7687.2010.01022.x.

Index